T0177232

THE LITTLE BOOK OF
WHALES

With color illustrations by Tugce Okay

ROBERT YOUNG AND
ANNALISA BERTA

PRINCETON UNIVERSITY PRESS
PRINCETON AND OXFORD

Published in 2024 by Princeton University Press
41 William Street, Princeton, New Jersey 08540
99 Banbury Road, Oxford OX2 6JX
press.princeton.edu

Library of Congress Control Number 2024931715
ISBN 978-0-691-26012-9
Ebook ISBN 978-0-691-26015-0

Typeset in Calluna and Futura PT

Printed and bound in China
1 3 5 7 9 10 8 6 4 2

British Library Cataloging-in-Publication Data is available

This book was conceived, designed, and produced by UniPress Books Limited

Publisher: Jason Hook
Managing editor: Slav Todorov
Creative director: Alex Coco
Project development and management: Ruth Patrick
Design and art direction: Lindsey Johns
Copy editor: Caroline West
Proofreader: Robin Pridy
Color illustrations: Tugce Okay
Line illustrations: Ian Durneen

IMAGE CREDITS:
Alamy Stock Photo: 17 Hemis; 59 ZUMA Press, Inc.; 100 Natalia
Pryanishnikova; 108 Volvox Inc.; 116 Pictorial Press Ltd.;
131 MET/BOT. **Dreamstime.com**: 23 Kajornyot. **Nature Picture Library**:
10 Tony Wu; 35 David Fleetham; 49 Brian Skerry; 55, 122 Doc White;
69 Hiroya Minakuchi; 70 Chris & Monique Fallows; 92 SCOTLAND:
The Big Picture; 96 Martha Holmes; 139 Doug Perrine; 144 Dave Watts.
Shutterstock: 24–5 Nataly23; 33 beaverboy56; 39t 3D-Horse; 39b Sebastian
Kaulitzki; 151 Phuttiwong. **Other**: 79 Callan Carpenter; 82 Ewa Krzyszczyk,
Shark Bay Dolphin Research Project. **Additional illustration references**:
12 Hans Thewissen; 21 Pavel Riha; 85 Tursiops; 105b Alessio Marrucci;
NOAA Fisheries (throughout).

Also available in this series:

CONTENTS

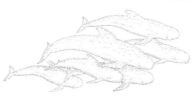

INTRODUCTION

Whales have long captured human interest. As the charismatic top predators in our oceans' ecosystems, these majestic sea mammals are as diverse as they are intelligent, from the Earth's largest living animal, the hundred-foot-long, two-hundred-ton blue whale, to the diminutive, less-than-five-foot-long, critically endangered vaquita.

TITANS OF THE WATERS

Evolving from land mammals, whales made the transition from land to water. Whales occupy a variety of environments, living in freshwater, estuarine and marine habitats. Huge distances are covered by highly migratory species such as the record-holding more than 10,000-mile (16,000-km) annual roundtrip made by the gray whale between birthing and calving in Mexican lagoons and feeding in polar latitudes. Others such as the sperm whale dive to depths of more than one-and-a-half miles for over two hours on a single breath-hold!

We catch only a brief glimpse of whales as they break the water's surface to breathe drawing air from the blowhole on the top of their heads. But thanks to developments in science and technology we are beginning to understand their complex social structures, extraordinary communication abilities, internal anatomy, behavioral patterns, and population sizes. High-tech tools include satellite tags and

microprocessors and drones for observing their movement patterns. Genomic studies of DNA variation allow identification and conservation studies while CT scans enable non-destructive three-dimensional (3D) reconstructions of organs and tissues, which is especially useful for exhibition and educational programs. As our fellow mammals living in the ocean, whales are sentinel species that provide insights not only into their fascinating biology but also into the health and resilience of coastal and oceanic ecosystems.

ABOUT THIS BOOK

Written for general animal enthusiasts and whale fans alike, this book reviews fascinating aspects of their biology, behavior, and conservation. Chapters 1 and 2 introduce whales, dolphins, and porpoises, presenting their diversity, distribution, and evolution 50 million years ago from the land to their present-day dispersal in all the world's oceans. Chapters 3 to 7 review their anatomical and physiological adaptations that enable them to live in the water, including breath-holding and diving, sensory systems, feeding, and sound and communication. Chapters 8 and 9 are focused on social organization, including mating and reproduction. Chapter 10 highlights various threats to cetaceans such as by-catch, pollutants, climate change, and noise, in addition to identifying conservation status, major international regulations, current issues, and the risk of extinction for various species. Chapter 11 examines dolphin myths and folklore, followed by Chapter 12 that highlights some curious facts about whales, including how they sleep, and mass strandings.

Robert Young and Annalisa Berta

WHAT IS A WHALE?

Found throughout the world's oceans, whales are mammals of the infraorder Cetacea and are easily recognized by their prominent tail fluke, a blowhole on top of their head, two front flippers, and no hind limbs. Cetaceans are the most diverse group within the marine mammals, a collection of unrelated, ocean-associated species that also includes the pinnipeds (seals, sea lions, and walrus), sirenians (manatees and dugongs), sea otters, and polar bears.

WHALE ORIGINS

Whales, like all marine mammals, evolved from land mammals and exhibit typical mammalian characteristics. They are homeothermic, or warm-blooded, and maintain a constant internal body temperature. Females bear live young, produce milk, and nurse their calves. Although they are mostly or entirely hairless as adults, they all have some hair for at least a portion of their lives. And even though they make their living primarily under water, cetaceans have lungs and breathe air.

Modern cetaceans are divided into two groups: toothed whales (Odontoceti) and baleen whales (Mysticeti). The latter have rows of filtering baleen plates instead of teeth. Although not visually obvious, the closest living relatives to cetaceans are the hoofed mammals (ungulates), specifically the even-toed ungulates (Artiodactyla). This group includes cattle, deer, pigs, and the like, most with familiar two-hooved tracks. Within the Artiodactyla, both molecular (DNA) and fossil evidence identify hippos as the closest relatives to whales. However, whales did not evolve from animals that looked like a hippopotamus. Both cetaceans and hippos evolved from a common mammalian ancestor that looked nothing like either of them.

WHALES, DOLPHINS, AND PORPOISES

In this book, the term "whale" refers to all cetaceans, including many smaller species that go by the name of dolphin or porpoise. There is no official size limit that distinguishes a whale from a dolphin or porpoise, and the common term of choice for a given species can vary depending on local traditions. Even the carefully defined taxonomic families of cetaceans become messy when it comes to common names. For example, the largest member of the dolphin family is the killer whale. Is a killer whale a big dolphin or is a dolphin a small whale? The answer to both questions, apparently, is yes.

← Baleen whales,
like this courting pair
of humpback whales
(*Megaptera novaeangliae
australis*), have rows of
baleen plates for filtering
plankton and fish.

THE FIRST CETACEANS

The "Age of Mammals" began with the extinction of the dinosaurs about 66 Mya (million years ago). The decline of large terrestrial predators and the abundance of unclaimed niches spurred a rapid diversification and evolution of mammalian species, and it was still early in this new age when the cetacean lineage diverged from the hippopotamid line, approximately 54 Mya.

A VERY TERRESTRIAL WHALE

In appearance, the "first whales" were clearly land mammals, and no one could have predicted that they were the first step in the extraordinary evolutionary journey toward the fully aquatic whales we know today. As it turns out, the rich fossil record that chronicles the transition from land mammal to modern whale is one of the most complete examples available of a dramatic macroevolutionary change. All early whales prior to the rise of modern toothed and baleen whales (about 34 Mya) are members of the Archaeoceti, and the very first of these in the fossil record were the Pakicetidae, a family of at least seven known species that lived in the early Eocene Epoch between about 53 and 49 Mya. The most complete fossils are of the genus *Pakicetus*, which was roughly the size of a wolf but with an elongated head and tail. It walked (and ran) on four limbs, with four toes on each hind limb and five digits on each forelimb.

↙ *Pakicetus* is the earliest known species of the cetacean line, linked to whales by its distinctively cetacean ear bone.

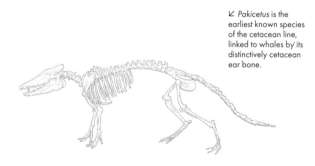

WHY IS *PAKICETUS* A CETACEAN?

Various skull and ear characteristics are indicative of cetaceans, but more than any other feature, the presence of an early cetacean ear bone, or tympanic bulla, classifies *Pakicetus* as an early whale. This is a distinctive structure that is only found in modern and fossil cetaceans (except for a concurrent sister group of *Pakicetus* with no known descendants). The tympanic bulla is a dense bone that encases the middle ear, and in all cetaceans it is substantially thickened on one side, an adaptation for improved underwater hearing. *Pakicetus* shares no obvious external traits with whales, but the first thought of a paleontologist who finds a *Pakicetus* cranium with its tympanic bulla intact would be: "This is an early cetacean!"

CLUES TO A SEMI-AQUATIC LIFESTYLE

Nearly all pakicetid fossils come from deposits in the hills of northern Pakistan, a region that at the time was a landscape of freshwater streams and pools in an otherwise hot and dry climate. A number of clues indicate that *Pakicetus* took advantage of this habitat to pursue a semi-aquatic lifestyle. Based on its small hands and feet, it was not a great swimmer. It also had heavy bones with less space for bone marrow, making it less buoyant in the water. This design is typical of wading mammals like hippos which run along the bottom while submerged. Its eyes were located close together toward the top of the skull, again common in submerged aquatic mammals that peer out at the surface. Finally, tooth wear patterns suggest *Pakicetus* ate fish, and analysis of stable isotopes in the tooth enamel indicates that its diet was mainly from a freshwater food chain.

THE AMPHIBIOUS WHALES

Following *Pakicetus*, several species-rich archaeocete families appeared between 49 and 47 Mya in the same Pakistan/India region. These amphibious whales were much better adapted to aquatic life but could still walk on land.

Ambulocetus natans, the "walking swimming whale," was a large predator in brackish and fresh waters, often compared to crocodilians in appearance and behavior, likely using its large hind flippers to burst forward to ambush prey. The lesser-known Remingtonocetidae were a long-snouted, giant otter-like family with small eyes, found mainly near murky river mouths. The Protocetidae were the most diverse and aquatic family, living in warm coastal ocean waters and using their hind flippers and tails to swim with a dolphin kick, although they likely did not have a tail fluke. Their vertebral design supported muscles for sustained swimming. They still walked on land periodically, although the pelvis of later species was no longer firmly connected to the spine, reducing or ending their ability to support their weight on land.

↓ Early cetaceans with legs are the only species that share the double-pulley ankle design of the Artiodactyla (their closest relatives).

↓ The four-toed hind foot of a protocetid, still with small, nail-like hoofs on the toes, were elongated and modified into webbed flippers.

→ Evolution of cetacean families. The amphibious whales (marked with *) experimented with various body plans. The protocetids were the only amphibious family to expand beyond their region of origin, eventually spreading to areas of Africa, Europe, and the Americas before disappearing 41 Mya. The Basilosauridae include (A) basilosaurus and (B) dorudontid.

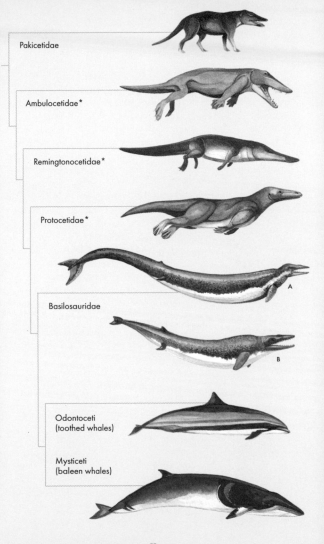

Pakicetidae

Ambulocetidae*

Remingtonocetidae*

Protocetidae*

Basilosauridae

A

B

Odontoceti
(toothed whales)

Mysticeti
(baleen whales)

THE FIRST FULLY AQUATIC WHALES

T he amphibious whales were ultimately replaced by the aquatic and very whale-like basilosaurs and dorudontids. Technically, the dorudontids are within the family Basilosauridae, but the two body plans differ and are often treated separately. The two species-rich groups coexisted from 43 to 34 Mya, and as the first whales to use a tail fluke for propulsion, they spread widely across the globe, although the basilosaurs were generally more coastal in distribution and did not venture into polar waters.

BASILOSAURS

The basilosaurs were the largest animals since the dinosaurs, reaching lengths of 60–65 ft (18–20 m). In fact, basilosaur means "king lizard," highlighting their misidentification as a giant reptile when first discovered. They still had hind legs and feet, but these were tiny and useless for swimming or movement on land, leading biologists to speculate that their primary use may have been to grasp mates during copulation. The structure of their vertebral column and tail vertebrae indicates they had a small tail fluke. They were somewhat long and slim for a whale, and their elongated and numerous vertebrae suggest they swam with a sinuous motion. Their forelimbs and hands likely formed pectoral flippers similar to those of modern whales and dolphins, and their nostrils were on top of their head but only about halfway back to the position of the blowhole on a modern whale. They had large heterodont teeth (meaning they had more than one type of tooth), which match the bite scars on dorudontid fossils, indicating that their smaller cetacean cousins were on the menu.

~ The heaviest basilosaur ~

A recently discovered partial skeleton of a basilosaurid from Peru is quite unique. Though similar in length to other large basilosaurs, its vertebrae and ribs were extremely thick and dense, presumably to provide extra ballast for bottom feeding in shallow coastal waters. Dubbed *Perucetus colossus*, its projected weight may rival that of a modern blue whale (*Balaenoptera musculus*).

DORUDONTIDS: LINK TO THE MODERN CETACEANS

Dorudontids would be mistaken today for a large dolphin, typically reaching lengths of 16–20 ft (5–6 m). Compared to basilosaurs, their tail fluke was proportionately larger, while their hind limbs were smaller and less complete, and also barely visible or potentially fully internal in some species. Otherwise, their pectoral flippers, nostril placement, and teeth were all similar to those of basilosaurs. Given their very similar body plan, dorudontids are believed to be the predecessors of modern cetaceans, while the basilosaur branch eventually went extinct. The world was heading for a big change in climate and ocean productivity, and cetaceans were going to change with it.

↑ The long vertebrae of a basilosaur skeleton "swim" through the sands of Egypt's Whale Valley, a UNESCO World Heritage Site.

RISE OF THE MODERN WHALES

Archaeocetes were replaced by the modern body design of the Odontoceti and Mysticeti beginning about 34 Mya. These changes were associated with a cooling global climate and the development of areas of high oceanic productivity.

CHANGING OCEAN CONDITIONS

Beginning about 34 Mya, the cooling of Antarctica was enhanced, in part, when Australia and South America tectonically separated far enough from Antarctica to allow the formation of a circumpolar current around the continent, isolating it from the warm currents that previously reached its shores. Subsequent cooling led to the formation of a polar ice cap at the South Pole, and the dense, cold, salty Antarctic water started to sink, establishing a new global circulation pattern. What goes down must come up, and the resultant increase in rising water, or upwelling, brought chemical nutrients (fertilizers) to the sunlit surface in predictable locations, producing enormous biological productivity. Some of the most productive upwelling zones occurred in the Southern Ocean where the first mysticetes evolved.

SKULL AND DENTITION CHANGES

The most dramatic cetacean developments involved changes to the skull and dentition. The nostrils, which had been moving back along the snout starting with the protocetids, finally reached their modern position, becoming a blowhole at the top of the skull. In doing so, many of the bones of the skull were altered and rearranged in a process called telescoping, which occurred very differently in the toothed and baleen whales.

~ Odontocetes ~

The odontocete skull was modified to support the muscles associated with sound generation and a fatty forehead, or melon, in front of the blowhole. These features are associated with echolocation, a unique sense that evolved in odontocetes during this period. Odontocete teeth changed as well. All archaeocetes were heterodont, but their dentition was slowly becoming less complex. With a few early exceptions, the odontocetes are homodont, meaning all their teeth are shaped in the same way, typically conical or peg-shaped. Finally, although odontocete skulls have two nares, or nostril holes, only one passage leads to the outside, forming a single blowhole (unlike mysticetes, which have two openings).

~ Mysticetes ~

Mysticetes modified their skulls by broadening their skull and palate to maximize the food-processing area of the mouth. These initial skull changes preceded the development of baleen, so the earliest mysticetes had just teeth, while some interesting transitional species had both teeth and baleen at the same time, and soon baleen had completely replaced teeth. Baleen is made of keratin (like fingernails and hair) and originates in the gum, generating rows of baleen plates that hang down from the upper jaw to filter plankton and fish from the water.

SHARED CHANGES

Two other modern features finally reached their culmination in odontocetes and mysticetes. The first of these is the complete absence of external hind limbs, although a remnant of the hind limb remains in modern cetaceans in the form of a small, vestigial thigh and/or pelvic bone sitting alone in the flank. The pectoral flippers also achieved their modern form, now with a fused, non-functioning elbow joint. Modern whales continue to reveal their terrestrial heritage, however, with each flipper containing all the bones of the arm, wrist, hand, and fingers.

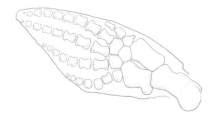

← A whale flipper contains all the bones of the vertebrate arm and hand, although their size and function are greatly modified.

EVOLUTION MARCHES ON

The cetacean evolutionary journey arrived at the basic design of modern toothed and baleen whales 34–30 Mya. This does not mean that nothing has really changed for whales in the last 30 million years. Indeed, whole families of strange and diverse whales have come and gone over that time.

Of the extant toothed whales, sperm whales were a diverse, early family to evolve, but only one species remains. The dolphin family, on the other hand, is now the largest odontocete family, but they only rose to prominence in the last five million years. The truly enormous baleen whale species have only evolved within that same time period, possibly in response to more intensive seasonal upwelling periods. Specifically, the modern blue whale can reach 100 ft (30 m) in length, and although shorter than some of the long-necked sauropod dinosaurs, it is the most massive animal ever known, weighing up to 160 tons. We are indeed living in large times.

↓ The extinct shark-toothed whales (Squalodontidae) were a successful early odontocete family the size of large dolphins.

They were the last group to maintain heterodont teeth (top), as compared to the homodont teeth seen in a modern killer whale (bottom).

→ Many variations to the basic odontocete and mysticete design have come and gone, ranging from this *Odobenocetops*, a whale with walrus-like tusks, to the enormous *Livyatan*, a truly intimidating 57-ft (17.5-m) sperm whale with foot-long (30-cm) teeth that preyed on other whales and large sharks.

WHALE TAXONOMY

Cetacean taxonomy has been suffering from an identity crisis. The term Cetacea was traditionally an order, equal in stature with other major mammalian groups such as carnivores, rodents, or primates. The Artiodactyla, or even-toed hoofed mammals, are also an established order, so the realization that cetaceans are nestled within a branch of the artiodactyls caused much confusion that has continued for over two decades

THE NEW ARTIODACTYLA ORDER

Some scholars have proposed a new order, the Cetartiodactyla, essentially giving equal stature to both groups. This is problematic because cetaceans diverged from just one of several branches of artiodactyls, so it is hard to justify equal stature. Others have argued to keep the two groups as separate orders for the sake of stability, even though they are clearly linked. Currently, the taxonomic committees for both the Society for Marine Mammalogy (SMM) and the American Society of Mammalogists (ASM) support cetaceans as an infraorder within the order Artiodactyla. This solution most accurately represents the taxonomic hierarchy, but it also means that a venerable order that traditionally included only hoofed mammals now quietly includes whales as one of its major subdivisions—an odd combination that is very confusing to the average observer. It appears no solution can please everyone.

~ Mysticeti and Odontoceti ~

The two major groups of cetaceans remain the Mysticeti and Odontoceti. Mysticetes have baleen instead of teeth, while the modern odontocetes are characterized by homodont teeth designed to grab and hold prey but not grind or chew. Thus, odontocetes generally eat their prey whole (although two orcas will sometimes cooperatively tear a large prey item apart). Odontocetes also have a fatty forehead, or melon, which assists in focusing sound for echolocation.

FORMAL AND INFORMAL GROUPINGS

When recognizing families and species, this book will follow the structure endorsed by the SMM Committee on Taxonomy. Under this structure, the mysticetes include fifteen species divided among four families and the odontocetes include seventy-nine species divided among ten families. These families are described in detail on the following pages

On a less formal note, nearly all mysticetes are commonly ascribed to the "great whales," a loosely defined group that includes cetaceans greater than about 42 ft (13 m) in length. With males reaching over 60 ft (18 m) in length, the sperm whale (*Physeter macrocephalus*) is the only odontocete typically counted among the great whales.

Identified by three ridges on the rostrum, the lunge-feeding Bryde's whale (*Balaenoptera edeni*) shown above is one of the great whales, reaching lengths of 50 ft (15 m).

BALEEN WHALE FAMILIES

Most baleen whales belong to the family Balaenopteridae, commonly called the rorqual whales. The ten rorqual species range in size from the giant blue whale (*Balaenoptera musculus*) at 100 ft (30 m) to the second smallest mysticete, the minke whale (*B. acutorostrata*) at 33 ft (10 m). Characterized by numerous throat pleats and a streamlined shape, rorqual whales feed by lunging forward, expanding their throat like a balloon to take an enormous gulp of water, and then filtering out prey through relatively short baleen as the water exits. The Balaenidae (four species) have pursued a simpler design and include the bowhead whale (*Balaena mysticetus*) and right whales. With long baleen and a strongly arched jawline, they simply open their mouth and swim, filtering zooplankton as they move slowly forward. Prioritizing girth over speed, they are robust whales and lack a dorsal (stabilizer) fin. The gray whale (*Eschrichtius robustus*) is the only member of the family Eschrichthiidae, as is the seldom seen pygmy right whale (*Caperea marginata*) in the family Neobalaenidae, the smallest of the mysticetes at 20 ft (6 m).

← Bowhead whale baleen can reach 10 ft (3 m) in length, hanging down from a highly arched upper jaw.

← Rorqual whales are more streamlined and fast with a broad, flattened skull and relatively short baleen.

↓ From left to right, representative members of the mysticete families: Balaenidae (southern right whale, *Eubalaena australis*),

Balaenopteridae (blue whale), Eschrichtiidae (gray whale), and Neobalaenidae (pygmy right whale). Species are drawn to scale.

TOOTHED WHALES: MARINE FAMILIES

T hree of the oldest extant families of toothed whales are the Physeteridae (sperm whale), Kogiidae (pygmy and dwarf sperm whales), and Ziphiidae (beaked whales). All three are deep divers, feeding primarily on squid and sharing a pattern of reduced dentition with teeth only in their lower jaws.

SPERM, PYGMY, AND DWARF SPERM WHALES

The sperm whale has a large, block-shaped head that is a quarter to a third of its body length, a modified melon called a spermaceti organ, a blowhole oddly placed at the front left corner of the head (producing a forward-angled blow), a narrow, underslung jaw, and a knuckled dorsal ridge instead of a dorsal fin. Pygmy sperm whales (*Kogia breviceps*) and dwarf sperm whales (*K. sima*) are smaller (less than 12 ft/3.6 m) and more dolphin-like, but their boxy head, spermaceti organ, and underslung jaw reveal their kinship with sperm whales.

THE MYSTERIOUS BEAKED WHALES

Beaked whales are the second largest cetacean family (containing 24 species), but they are unfamiliar to most people, given their preference for deep waters of the shelf edge and open ocean, and their reclusive behavior. They are whale-sized (13–43 ft/4–13 m), but have a dolphin-like rostrum (snout-like projection) and two expandable, V-shaped throat grooves. In most species, females have no teeth and males only have one or two pairs of often large teeth in the lower jaw, which erupt during puberty.

A few species are well studied: northern bottlenose whales (*Hyperoodon ampullatus*) and Cuvier's beaked whales (*Ziphius cavirostris*) were targeted by whalers in the previous century, and researchers have identified several beaked whales as among the deepest diving marine mammals. Many species, however, are quite mysterious, including some Southern Ocean species with no confirmed live sightings. For more than a century, the spade-toothed whale (*Mesoplodon traversii*) was known from only a few collected skulls until a mother–calf pair stranded in New Zealand in 2010.

DOLPHINS AND THEIR RELATIVES

The oceanic dolphins and their relatives make up three well-known families: the Delphinidae (dolphins), Phocoenidae (porpoises), and Monodontidae (belugas and narwhals). Delphinidae is the largest cetacean family, with nearly 40 species. United by subtle skull characteristics and a full set of conical teeth, they include many familiar widespread species, including bottlenose dolphins (*Tursiops* species), common dolphins (*Delphinus delphis*), and spinner dolphins (*Stenella longirostris*), as well as orcas, or killer whales (*Ocinus orca*), and pilot whales (*Globicephala* species).

~ Narwhals and belugas ~

Few realize that the narwhal (*Monodon monoceros*) and beluga whale (*Delphinapterus leucas*) have nearly the same body plan, including body lengths up to 16 ft (5 m); a large, rounded melon; relatively small, rounded tail flukes and flippers; a more flexible neck than most cetaceans; and a dorsal ridge rather than a dorsal fin. Narwhals have only two teeth, one of which, in males, erupts from the upper jaw, forming the iconic, long (up to 10 ft/3 m), spiraling tusk as a secondary sexual characteristic. The distinctively all-white belugas have more typical dentition, each with 30–40 conical teeth.

~ Dolphins and porpoises ~

The difference between dolphins and porpoises is a common question. These two terms are often used interchangeably for local common names, but to a cetacean biologist, the porpoise family specifically refers to seven to eight species of generally small odontocetes ($4^1/2$–$7^1/2$ ft/1.4–2.3 m) with a blunt rostrum and distinctive, spade-shaped teeth. Collectively, the porpoises are widespread, ranging from polar to tropical latitudes and from oceans to estuaries and rivers.

→ With its small, underslung jaw and false gill mark, the dwarf sperm whale projects a shark-like appearance.

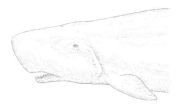

TOOTHED WHALES: RIVER DOLPHINS

Only the river dolphins remain, with five species spread across four families. Each family evolved separately, but they all converged on a similar body plan, including a long, narrow rostrum with many spiky teeth, reduced eye orbits, and a flexible neck

Three families are related to oceanic dolphins and include the Amazon river dolphin (*Inia geoffrensis*) in the Iniidae family, found throughout the Amazon and its tributaries; the small franciscana (*Pontoporia blainvillei*) in the Pontoporiidae family, found in coastal and estuarine waters of Argentina, Uruguay, and southern Brazil; and the Yangtze river dolphin (*Lipotes vexillifer*) in the Lipotidae family, in China, which was last sighted in 2004 and presumed extinct from fishing bycatch and dam-related habitat fragmentation. An older family (Platanistidae), related to sperm whales and beaked whales, includes two additional species, the Ganges river dolphin (*Platanista gangetica*), found in India, Bangledesh, and Nepal, and the Indus river dolphin (*P. minor*), which is found mainly in Pakistan. The eyes of these two species do not have an eye lens, so they are functionally blind and navigate using echolocation.

↓ Two tusks erupt from the lower jaw of the male Blainville's beaked whale (*Mesoplodon densirostris*) at puberty. They are its only teeth.

↓ Ganges and Indus river dolphins have a unique dorsal skull crest, believed to direct echolocation sounds forward through the melon.

→ Representative odontocetes of the families Physeteridae (sperm whale), Kogiidae (pygmy sperm whale), Ziphiidae (Blainville's beaked whale), Delphinidae (common dolphin), Phocoenidae (Dall's porpoise, *Phocoenoides dalli*), Monodontidae (beluga whale), and Platanistidae (Ganges river dolphin). Species are not drawn to scale.

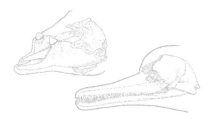

BIOGEOGRAPHY

The distribution of whales, dolphins, and porpoises is driven by spatial and seasonal patterns of ocean productivity. With their high mammalian metabolism, cetaceans must go where prey is abundant. This is especially true for baleen whales, whose feeding technique requires dense aggregations of prey.

Ocean productivity depends on phytoplankton, the microscopic algae that form the base of the marine food chain. To survive, phytoplankton require light, available only in near-surface waters, and dissolved nutrients such as nitrate and phosphate, which are fertilizers required for photosynthesis. Surface nutrients are sparse, as they are rapidly taken up by phytoplankton, enter the food chain, and eventually become entrained in marine detritus that sinks out of the surface waters. As these particles break down, the nutrients are returned to the dark, deeper waters. The most productive regions in the ocean therefore occur where nutrient-rich deep water is returned to the sunlit surface through wind-driven mixing and upwelling.

PRODUCTIVE WATERS

Though somewhat counterintuitive, most of the tropical and subtropical ocean is unproductive due to warm, low-density surface waters that overlie colder, dense waters, separated by a thermocline and forming a strongly layered ocean that blocks the recirculation of nutrients. The most productive waters occur in cold subpolar and polar waters, where the lack of this warm, shallow layer allows for mixing and abundant surface nutrients all year round. Even here, the productive season centers only around summer, when daylength is long in high latitudes and sunlight is overabundant.

Many baleen whale species base their movements on these latitudinal and seasonal patterns, migrating to productive high latitudes to feed in summer and then moving into less productive warmer waters, often fasting for long periods, to give birth and mate in winter (see Chapter 9, page 112). Examples of these impressive migrations include the gray whale, which migrates between summer feeding grounds in the Bering Sea (North Pacific) and winter calving

CATEGORIES OF DISTRIBUTION

Only a few whale species, such as sperm whales and killer whales, are truly cosmopolitan in their distribution, found in all oceans and all latitudes, from the Equator to the edge of the polar ice. A number of dolphin and small whale species are nearly as widespread, although they tend to avoid true polar and subpolar waters. A few cetaceans, including the bowhead whale and the beluga and narwhal, are classified as circumpolar, remaining in polar and subpolar waters all year round. And finally, many species are endemic to a localized area, such as the river dolphins, several species of dolphins and porpoises, and even one baleen whale, the recently classified Rice's whale (*Balaenoptera ricei*), see page 32.

grounds along the western coast of Mexico, and various humpback whale (*Megaptera novaeangliae*) populations migrating between the Bering Sea and Hawai'i, or coastal Antarctica and Fiji, or southern Greenland and the Caribbean Sea.

On top of this general latitudinal and seasonal pattern, specific areas of the ocean have highly productive, wind-driven upwelling zones that bring nutrient-rich bottom waters to the surface. These regions are most notable in waters surrounding Antarctica, along areas of the west coast of the Americas and Africa, and in the northwest Indian Ocean. In near-coastal and estuarine regions, productivity is often enhanced by nutrient additions from rivers and from efficient mixing in shallow waters. All of these are prime areas to find cetaceans.

→ Productive coastal upwelling occurs when winds (A) drive surface waters offshore (B), causing rising bottom waters (C) to bring nutrients (fertilizers) to the surface.

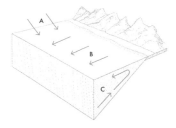

2. Diversity and Distribution

NEW SPECIES, SUBSPECIES, AND HYBRIDS

The discovery of new cetacean species is not unusual, though the term "discovery" is misleading. These new species have been hiding in plain sight, classified as members of established species until careful studies reveal morphological and genetic differences that justify reclassification.

Two recent examples are Rice's whale and Tamanend's bottlenose dolphin (*Tursiops erebennus*), described in 2021 and 2022, respectively. Rice's whale was formerly classified as a Bryde's whale, a rorqual whale unique among mysticetes for its global distribution in tropical and warm temperate waters. The new species is endemic to the northeastern Gulf of Mexico and is highly endangered, with an estimated population of fewer than 50 individuals. The Tamanend's bottlenose dolphin is a dramatic reclassification of the highly studied bottlenose dolphins along the coastal waters and estuaries of the United States east coast.

CLASSIFICATION GUIDELINES

Oddly enough, there is no universal agreement about what defines a species, subspecies, or hybrid. Traditionally, only members of the same species can mate and produce fertile offspring, while subspecies are groups within a species with distinct features and genetics, and hybrids are the infertile offspring of two closely related species. In application, however, these definitions often break down, so researchers have identified guidelines for determination, based on multiple lines of evidence.

Many baleen whales, for example, have separate populations in the Northern and Southern hemispheres. In right whales, for example, the identified differences justify separate species for the North Atlantic, North Pacific, and Southern right whales (*Eubalaena glacialis*, *E. japonica*, and *E. australis*, respectively). However, in blue whales, fin whales (*Balaenoptera physalus*), humpback whales, and sei whales (*B. borealis*), the northern and southern populations have been classified as subspecies, rather than separate species.

SUBSPECIES AND HYBRIDS

Subspecies have been defined for numerous dolphin and porpoise species, and many hybrids have been observed in captivity and the wild. Bottlenose dolphins hybridize with at least ten species of delphinids, and even blue whale–fin whale hybrids and beluga–narwhal hybrids have been observed. Sea Life Park in Hawai'i is famous for two generations of "wholphins," common bottlenose dolphin–false killer whale hybrids (*Tursiops truncatus* and *Pseudorca crassidens*, respectively). The original wholphin, a female, grew up and successfully mated with a bottlenose dolphin, demonstrating she was fertile. There goes the traditional species definition! Although vexing for taxonomists, the whales and dolphins undoubtedly don't give a flipper about this.

↑ This "wholphin" (left) is the offspring of a false killer whale (*Pseudorca crassidens*), with a rounded, blunt rostrum, and a common bottlenose dolphin (right).

A MAMMAL IN THE SEA

A truly fascinating aspect of whales is the anatomical and physiological adaptations they have made, as mammals, to transition from a terrestrial to an oceanic existence. This chapter will focus on cetacean solutions to challenges involving locomotion, managing salt and water balance while living in salt water (osmoregulation), and maintaining the elevated, constant body temperature required of mammals (thermoregulation). Other interesting physiological challenges include diving and breath-holding and the function of sensory systems under water, but those will each get a chapter of their own.

LOCOMOTION

Mammals are dominated by walkers, runners, climbers, and even fliers (bats), but marine mammals have transformed into swimmers. Seawater is about 800 times more dense than air, which sounds like an enormous additional burden, but there is an upside to this condition. The trick is to develop a hydrodynamic shape to minimize drag and then establish some buoyancy control to exploit the inherent support of a dense medium. Once these abilities are in place, whales can take advantage of gliding and momentum to move though the water with great efficiency.

SALT AND WATER BALANCE

Ocean water is much saltier than mammalian body fluids, so salts will passively diffuse from high to low concentration into the body across permeable membranes like the gut, displacing water, which will passively diffuse out. Humans, of course, cannot live without fresh water—in fact, drinking salt water causes rapid dehydration and, if continued, death. Cetaceans, on the other hand, have evolved osmoregulatory mechanisms to excrete accumulated salts and retain fresh water. For more on this, see pages 40–41.

THERMOREGULATION

The fundamental thermoregulatory challenge for whales is that thermal conductivity in water is 25 times greater than it is in air, or in other words, heat is removed from the body 25 times faster in water. A human can sit comfortably all day in a 72°F (22°C) room but we are shivering in half an hour if we sit in water of the same temperature. In contrast, a narwhal (*Monodon monoceros*) lives in icy-cold waters year-round, while a deep-diving sperm whale (*Physeter macrocephalus*) at the Equator may move quickly from 86°F (30°C) water at the surface to 39°F (4°C) water at the bottom, and remain in those cold waters for an hour or more. For more on this, see pages 42–43.

↑ Though apparently attempting to fly, this pantropical spotted dolphin (*Stenella attenuata*) is perfectly adapted for life in the ocean.

THE HYDRODYNAMIC SHAPE

The quest for hydrodynamic efficiency begins with body shape. Whales have eliminated unneeded appendages and protrusions that may cause turbulence, while approximating a classic "fusiform" shape: tapered at the front, reaching maximum girth about one-third of the way back, and then gently tapering toward the tail. This carefully crafted shape is maintained against the relentless pressure of pushing through dense water by a subdermal sheath of connective tissue, which acts as elastic scaffolding connecting muscles, tendons, bones, and blubber, holding them all in position like a pressurized cylinder. Faster swimmers usually have a dorsal fin to stabilize against rolling.

A SMOOTH SURFACE

The surface texture of the skin also assists in swimming efficiency. Except for some hairs on the rostrum or chin, the body of cetaceans is hairless, smooth, and somewhat rubbery in texture. In some dolphins, barely perceptible ridges in the skin help to maintain the parallel flow of water along the surface of the body. The thick skin has fingerlike projections connecting the epidermis and dermis (outer and inner layers). These provide a large surface area for skin cell production, resulting in a very high rate of turnover for the outer layer of skin, which helps to maintain a smooth surface as cells slough off and limit the number of "fouling" organisms that can potentially attach to the skin and increase drag. Nonetheless, a number of specialized barnacle species have adapted their attachment mechanisms and successfully colonized cetaceans.

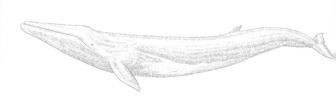

STREAMLINING THE BODY

Body parts that stick out have been internalized or streamlined. The hind limbs have been eliminated and the forelimbs modified into compact flippers for steering and adjusting pitch up and down. Whales lack a clavicle, or collar bone, eliminating turbulence from shoulders. In males, the testes are internally located within the body cavity and the penis is retracted within a genital slit, while in females the mammary glands and nipples are located beneath paired mammary slits. Beaked whales have flipper pockets, depressions in their body for their flippers to lay into for a streamlined profile when gliding. For all species, the strategic use of blubber can smooth out any remaining bumps or imperfections.

POWERING THE STROKE

Forward momentum is powered by massive muscle bands running the length of the body that beat the large tail fluke up and down. The tail stalk, or peduncle, is laterally flattened, helping to streamline this motion. The fluke is a large hydrofoil that pivots up and down as the tail beats, directing powerful sweeps of water on both the up and down kicks. The wide flukes are formed by a network of stiff connective tissue, and have no bones other than the final tail vertebrae that taper into the central base. Humans and most land animals swim with less than 50 percent efficiency, because half of the stroke motion is a recovery to get in position for the next pull, but the hydrofoil or lift-based system of whales (and fish) can be up to 80 percent efficient. Killer whales can achieve burst speeds of over 30 mph (50 km/h), and some dolphin species and even the long and streamlined fin whale (*Balaenoptera physalus*) are nearly as fast.

← With its streamlined shape and large size (up to 90 ft/27 m), the fin whale may be the fastest baleen whale.

MUSCULOSKELETAL SYSTEM

With no pelvis or hind limbs, the long vertebral column dominates the cetacean skeleton. Along nearly the entire length, the vertebrae are distinctively shaped with prominent dorsal and lateral extensions (processes) and ventral extensions (chevron bones) on the tail vertebrae, which provide muscle attachment sites and leverage for the massive muscle bands that run the length of the body. Alternating contractions of epaxial (upper band) and hypaxial (lower band) muscles cause the tail to beat up and down. Overlapping notches of the thoracic (chest) vertebrae and adjacent lumbar vertebrae provide stability to this central region, acting as an anchored base for the muscles to maximize the power of the dolphin kick at the far end of the tail. The seven cervical, or neck, vertebrae are compressed and stacked together in most species, with one or more fused together. This results in a short, inflexible neck that stabilizes the head and provides hydrodynamic stability while swimming.

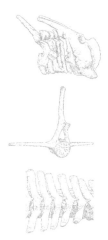

← Close up of compressed dolphin cervical (neck) vertebrae that provide stability. Some odontocetes, such as river dolphins, have more flexible necks at the cost of swimming speed.

↙ A single vertebrae, showing dorsal and transverse processes, and a set of thoracic (chest) vertebrae with overlapping notches that provide stability to that region of the spine.

→ The customized vertebral column and axial muscles work together to produce remarkable swimming power. The epaxial (A) and hypaxial (B) muscles attach to the many dorsal extensions and wings (processes) of the vertebrae, as well as the ventral extensions of the chevron bones (C) toward the tail, providing leverage for muscle contraction as the tail fluke beats up and down.

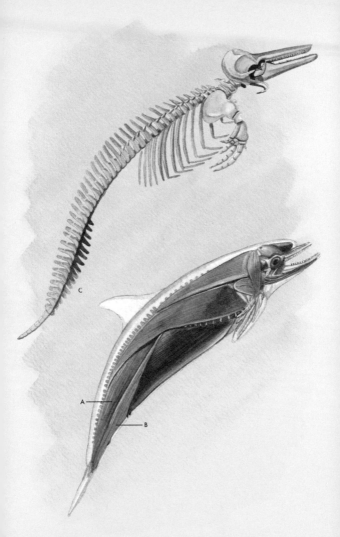

39

SALT AND WATER BALANCE

Humans are largely unaffected by short-term submersion in salt water, but if a human drinks salt water, dissolved salts will diffuse across the gut into the blood, where they will soon be removed by the kidneys. Unfortunately, the human kidney cannot produce highly concentrated urine, so a greater volume of water than was originally ingested is required to excrete the absorbed salts, causing rapid dehydration.

THE CETACEAN SOLUTION

Whales do two things differently: they get their water primarily from their food and they have kidneys that can make a more concentrated urine. Most bony fish actively regulate their body fluids to contain about one-third the salt concentration of seawater, similar to the body fluids of the whale. Thus, a whale that eats fish gains fluids (water) without excess salts. Whales that eat plankton or squid do not get the same benefit as fish-eaters because invertebrate bodies are isotonic with seawater (they have the same salinity), providing little advantage over drinking seawater.

Even eating fish is not a perfect solution, as an end product of digesting fish protein is urea, a nitrogenous waste that must be excreted in the urine, again leading to water loss. Thus, obtaining water from

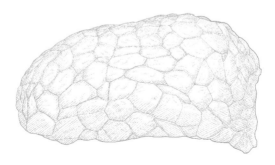

food is not a complete solution. Ultimately, cetaceans rely on more efficient kidneys made up of many lobes, or reniculi, structures that are roughly the size and shape of grapes. Each reniculi acts as a mini-kidney, filtering the blood and collecting wastes to direct them toward the bladder. The cumulative efforts of the many reniculi— up to thousands per kidney—are more efficient than a typical kidney and produce urine with a higher salt concentration than seawater. Thus, cetaceans can excrete their accumulated salts without a net loss of water.

The impressive capabilities of marine mammal kidneys, however, do not make them the champions of salt and water balance among mammals. That title goes to desert-dwelling mammals, such as the kangaroo rat, whose kidneys can produce urine that is 17 times more concentrated than their blood, retaining fresh water while excreting unwanted wastes.

~ What about freshwater species? ~

River dolphins live in fresh water and do not have the same issue with salinity. It seems that other cetaceans should also be able to invade fresh waters easily, and although this is true in the short term, even estuarine cetaceans can experience potentially lethal difficulties with prolonged exposure to fresh water. The skin of most cetaceans is adapted to salt water, and skin lesions, infections, and increased vulnerability to unfamiliar freshwater pathogens can develop with extended exposure.

← The cetacean kidney is made up of many functional mini-kidneys called reniculi—up to thousands per kidney, depending on the species.

THERMOREGULATION

Cetaceans maintain an internal body temperature similar to that of humans and most land mammals. This is a challenge since water draws heat away much faster than air. One potential solution is to simply generate more heat with a higher metabolic rate, and in fact, most cetacean species have a slightly elevated metabolism for their size compared to land mammals. There is a limit to this strategy, however, because a higher metabolism requires more fuel, and whales can't simply buy more food at the grocery store. A larger size is also an advantage because in larger animals the ratio of the body's surface area to its volume is small, meaning that a large volume of tissue is metabolizing and generating heat, but there is proportionately less surface area through which the heat can escape. This is an obvious advantage for a large whale, but less so for a small porpoise. Nonetheless, many porpoises live quite successfully in cold waters, so there must be additional strategies.

CIRCULATORY ADAPTATIONS

The circulatory system is also adapted to retain heat through an ingenious system called countercurrent heat exchange. When blood runs near the surface of the body, it releases heat into the water and cools down. As that cold blood returns, the vein is routed alongside a countercurrent artery with warm blood coming from the body's

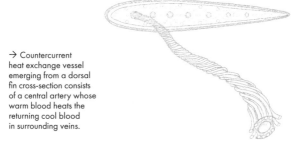

→ Countercurrent heat exchange vessel emerging from a dorsal fin cross-section consists of a central artery whose warm blood heats the returning cool blood in surrounding veins.

core. Because the blood in the vein and artery are flowing in opposite directions, there is a diffusion gradient for heat along an extended length of the vessels (the artery is always a little warmer than the vein, allowing heat to flow from high to low). Thus, the incoming blood is quickly warmed up to the desired temperature before returning to the body's core. To say that the two vessels run alongside one another is an oversimplification. In fact, the vein surrounds the artery, like a tube within a tube, although the outer "tube" is typically structured as multiple veins surrounding the central artery. This combined structure of countercurrent vessels maximizes the contact surface area and efficiency of heat exchange. Countercurrent vessels are prominently located in the areas most vulnerable to heat loss, including thin appendages such as the dorsal fin, tail fluke, and flippers, as well as the eyes and tongue (a mysticete tongue is enormous and is frequently exposed to cold water).

COOLING DOWN

Often forgotten, thermoregulation also requires cooling mechanisms when activity levels increase. In a dolphin dorsal fin, for example, countercurrent vessels run through the center of the fin, away from the cold edges. As activity heats up, the heart rate increases, pumping more blood and dilating the arteries. When the central artery of a countercurrent vessel expands, the surrounding veins are compressed, blocking the returning blood flow. The blood is diverted to peripheral veins near the surface of the fin, where heat can dissipate into cool adjacent waters. This elegant solution automatically switches from a warming to a cooling strategy when heat generation increases.

Another cooling example is the male testes, which require a cooler temperature than the rest of the body for sperm production. Most mammals solve this with external, scrotal testes, but cetacean testes are internal. They are cooled by specialized blood vessels carrying chilled blood from the dorsal fin and tail fluke that have bypassed any countercurrent warming.

BLUBBER

All cetaceans rely on an insulating layer of blubber surrounding their bodies between the skin and muscle. Blubber is a unique tissue not found in land mammals, consisting of a mixture of lipids (fats) and fibrous connective tissue. Oily yet firm and springy in texture, it serves multiple functions, including insulation (low thermal conductivity), energy and water storage for times of fasting or low food intake, buoyancy, and streamlining of the body. Both the thickness and composition of the blubber can vary by species and by position on the body. A higher ratio of lipid to connective tissue provides better insulation, and heat conductivity also varies depending on the type of lipid. Blubber thicknesses can range from less than an inch (2.5 cm) in some dolphins and porpoises to 19 in (48 cm) in bowhead whales. The thickness often changes seasonally in response to both changing environmental temperatures and extended periods of fasting or feeding.

↓ The blubber layer of Arctic bowhead whales is the thickest of any species of whale and can be 19 in (48 cm) thick and greater than 40 percent of the whale's total body mass.

↓ Exposed to frigid polar waters for long periods of time while skim feeding, the enormous tongues of bowhead whales rely on numerous countercurrent vessels to retain heat.

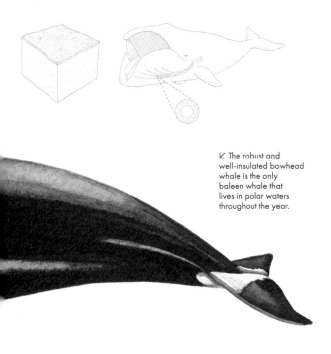

↙ The robust and well-insulated bowhead whale is the only baleen whale that lives in polar waters throughout the year.

DIVING ABILITIES

With a maximum recorded dive depth of 1.86 miles (2,992 m) and a duration of 3.7 hours, the Cuvier's beaked whale (*Ziphius cavirostris*) holds the record as the deepest diving air-breathing animal. The much shallower-diving loggerhead sea turtle has the longest dive duration, with winter dives lasting 10.2 hours. Unlike the cold-blooded loggerhead, however, whales have a higher mammalian metabolism.

Diving abilities vary greatly among whales. The deepest divers, which also include sperm whales (*Physeter macrocephalus*) and various beaked whales, are all squid specialists. Squid and their relatives are essentially the only deepwater prey worth the trip. Most dives by these diving specialists are shorter and shallower than the record-setters. Numerous species with intermediate abilities exploit more abundant prey resources at depths of around 1,600–3,300 ft (500–1,000 m) or less, typically making dives less than 20–30 minutes long. Meanwhile, many coastal and river dolphins never encounter deep water and only dive for a few minutes, although most can stay down longer if needed.

↓ Time-depth recorder tags, like this suction cup D-tag, attach to whales for anywhere between hours and days. They collect detailed diving data for analysis.

↓ Bowhead whales (*Balaena mysticetus*) are the deepest diving mysticetes (1,909 ft/ 582 m), feeding on zooplankton overwintering near the seabed.

→ The Cuvier's beaked whale, inhabiting deep ocean waters and reaching lengths of up to 23 ft (7 m), is the deepest diving air-breathing animal. Only three other species are known to dive over 6,562 ft (2,000 m) deep, the sperm whale, the northern bottlenose whale (*Hyperoodon ampullatus*), and the southern elephant seal.

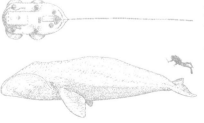

OXYGEN EXCHANGE

Oxygen storage begins with oxygen exchange. Since whales surface less frequently than a typical mammal takes a breath, they must exchange more air per breath. In fact, whales exchange 80–90 percent of the air in their lungs with each breath, as compared to about 10 percent for humans.

Not only do the lungs need to exchange a large volume of air, but they also need to do so quickly. The dorsal position of the blowhole on top of the head greatly improves the swimming efficiency of cetaceans when breathing, but whales need to exhale and inhale quickly in order to remain in an optimum swimming posture, minimize surface turbulence and drag, and maintain momentum. The powerful cetacean diaphragm sits at an oblique angle within the chest, and when it contracts it squeezes the lungs against the solid dorsal wall of the body cavity, virtually crushing the lungs as the air is forced out. The elastic lung tissue is able to withstand this abuse and still spring back when air rushes back in.

LUNGS ARE NOT FOR STORAGE

An understandable assumption is that whales have enormous lungs to store more oxygen, but in reality, that is not the case. In addition to being a buoyancy problem, oxygen stored in the lungs is oxygen not being used, and the better strategy is to redistribute it to where it is needed. Once air fills the lungs, the oxygen is quickly transferred to the blood via dense capillary networks that surround the terminal air sacs, or alveoli.

THAR SHE BLOWS!

The iconic whale blow (a right whale is shown above, with its spaced nostrils producing a distinctive V-shaped blow) occurs as a muscular plug contracts to open the blowhole and air is rapidly exhaled, an event that is vastly underappreciated. A blue whale (*Balaenoptera musculus*), for example, exhales about 528 gallons (2,000 liters) of air (picture one thousand 68-fl-oz/2-liter drink bottles) in just a couple of seconds through two blowholes about 18 in (46 cm) in diameter. The warm air is further heated under this pressure as it moves through airways toward the blowhole and then instantly expands and cools after it escapes, causing water vapor in the breath to condense. This sudden appearance of water is often mistaken for a waterspout coming out of their head.

OXYGEN STORAGE

Oxygen storage strategies vary by whale species and diving ability. While humans store at least half of their oxygen in their lungs, that proportion drops to about a third in shallow-diving coastal bottlenose dolphins (*Tursiops* species) and to only about 5 percent in deep-diving sperm whales. Instead of the lungs, cetaceans store oxygen primarily in their muscles and blood. In muscle, oxygen binds to myoglobin, a protein also present in the muscle of land mammals, but its concentration in whales is 10–30 times higher, especially in the swimming muscles.

In all mammals, oxygen in the blood binds to hemoglobin, a protein found within red blood cells, but cetaceans have a higher concentration of red blood cells per unit of blood, and the volume of blood in their body is two to three times the expected volume for a similar-sized land mammal, especially in deep-diving species. Collectively, these adaptations result in a stored volume of accessible oxygen that can be several times more than what would be expected for land mammals.

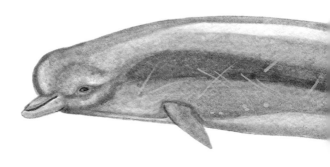

← A comparison of oxygen storage sites for a deep-diving sperm whale, a shallow-diving coastal bottlenose dolphin, and a human, showing the proportions stored in the blood (dark), lungs (medium), and muscle (light). Each also stores a small amount of oxygen directly in various organs, especially the brain, which is not shown in this comparison.

↓ The northern bottlenose whale, which is found in the North Atlantic, is among the deepest-diving beaked whales, with dives of over 6,560 ft (2,000 m) deep and durations of more than two hours. The deepest-diving whales store more oxygen in their blood and muscles and less in their lungs.

OXYGEN RATIONING

Additional oxygen storage by itself would enable some whales to dive up to three times longer than a really good human free-diver, but that is a long way from a two-hour sperm whale dive. Oxygen use is behaviorally reduced by the liberal use of passive gliding, especially during the descent and ascent, but the success of long dives relies much more on a series of circulatory system adaptations that ensure the frugal and efficient use of oxygen.

BRADYCARDIA

The first of these adaptations is a dramatic decrease in heart rate, or bradycardia, that occurs during a dive. The heart rate drops to about 10 percent of its normal rate, although there appears to be some degree of conscious control, as longer dives have more extreme bradycardia from the very start. Bradycardia causes oxygen rationing: slower delivery means slower usage.

SELECTIVE RESTRICTION

The second major adaptation is peripheral vasoconstriction, a systemic constriction of blood vessels that cuts off most circulation to the muscles and peripheral areas of the body. The muscles already have their own oxygen stores in their myoglobin, so vasoconstriction prioritizes circulating oxygen for important organs like the heart and brain. Other organ systems, such as the liver, kidneys, and intestines, receive reduced blood flow and decrease their activity during diving.

↘ Bradycardia champions: the heart rate of diving narwhals (*Monodon monoceros*) can drop to only three beats per minute.

CARBON DIOXIDE BUILDUP

Carbon dioxide is the by-product of cellular respiration and, as a dive progresses, it continues to build up in the blood and tissues, increasing acidity. In humans, a high carbon dioxide concentration in the blood triggers the eventually irresistible breathing reflex, but cetaceans are voluntary breathers, so they must think about every breath they take. Whales are resistant to this impulse to breathe and are more tolerant of increased acidity, in part due to greater buffering capacity that limits the change in pH. Nonetheless, when whales finally surface after a dive, they switch to a very rapid heart rate for a short time (tachycardia) in order to rapidly circulate and remove the carbon dioxide (which is exhaled from the lungs), restore blood pH, and recharge the oxygen stores in the blood and muscles.

~ Temporary blood storage ~

Such extreme vasoconstriction means that the volume of blood in the body now has much less plumbing to reside in, and without further adjustments, blood pressure would spike and the heart and vessels would burst. To prevent that from occurring, cetaceans have extensive areas of retia mirabilia (wonderful nets), tangled complexes of normally empty blood vessels that are able to receive blood and act as overflow reservoirs when other vessels have closed down. The most prominent of these networks is ingeniously located adjacent to the lungs along the dorsal body cavity wall, and when increasing pressure during dives compresses the air in the lungs, the increased volume of the blood storage vessels helps to partially counter the collapse of this space.

~ Pumping against pressure ~

Many whales also have an enlarged aortic bulb, which is a wide section of the main artery leaving the heart. The bulb's elastic walls swell with blood when the heart beats, and then the elastic recoil keeps the blood moving during the time between beats, helping to mediate blood pressure fluctuations and relieve strain on the heart.

ANAEROBIC RESPIRATION

E ven with their many adaptations for storing and using oxygen, whales will eventually run out of oxygen if they stay submerged long enough. That does not always mean the end of a dive, however, since whales can extend their dive time using anaerobic respiration.

WHAT IS ANAEROBIC RESPIRATION?

The longest dives require anaerobic respiration, or the conversion of carbohydrates into energy in the absence of oxygen. This is a short-term solution to oxygen depletion that is employed by all mammals, but it has a cost. Aerobic (oxygenated) respiration is 19 times more efficient than anaerobic respiration at converting sugars into energy, and the by-product of anaerobic respiration is lactic acid, which must be broken down and removed fairly quickly. That process requires oxygen, so the animal cannot long delay its return to oxygenated conditions. Thus, anaerobic respiration is only used when it is worth it—for example, to remain in a rich prey patch during a dive.

HIGH TOLERANCE FOR ANAEROBIC RESPIRATION

Whales and deep-diving seals are experts at managing anaerobic respiration, since their blood and tissues have a high buffering capacity to mediate increases in acidity (see page 53). In addition, peripheral vasoconstriction during the dive ensures that most of the lactic acid produced in the muscles and other isolated tissues does not circulate in large amounts, thus protecting the brain and heart from acidic conditions. Upon surfacing from an anaerobic dive, whales must circulate fresh oxygen to metabolize and eliminate the lactic acid. This is enhanced by the increased heart rate that always occurs upon surfacing, but for anaerobic dives, the heart rate is particularly fast.

→ A sperm whale descends to the depths. On their longest dives, more than half of the dive is anaerobic.

HOW MANY DIVES GO ANAEROBIC?

It is difficult to determine how long a whale's oxygen supplies will last. Sperm whales can dive for two hours, but most dives last less than 45 minutes, and it is assumed that the aerobic dive limit is about that long. Whales often take a longer surface interval between dives following an anaerobic dive, giving more time to eliminate the lactic acid and recharge oxygen stores. However, this is not always the case, and beaked whales in particular are mysterious. Cuvier's beaked whales do not take longer breaks after dives exceeding 75 minutes, suggesting that they either have very long aerobic dive limits or unusual abilities to manage the effects of anaerobic respiration.

PRESSURE

By simply diving in the deep end of a swimming pool, we are acutely aware of the pressure from the weight of the overlying water. Whales dive much deeper and have developed adaptations to deal with increased pressure.

THE COLLAPSIBLE RIB CAGE

You have likely felt the pain of an ear squeeze when diving just a few meters deep. Whales do not have that problem because they don't have large air spaces in their middle ears or sinuses to be squeezed. Their lungs, however, compress as water pressure increases. If a human free-diver goes too deep, their lungs and chest will collapse beyond recovery. In contrast, the more elastic fibers of cetacean lungs spring back from complete collapse, and interestingly, so can their rib cage. Whales have more floating ribs (ribs that do not connect to the sternum) than terrestrial mammals, and even the sternal ribs are either cartilaginous or connect with flexible ligaments that allow the rib cage to temporarily collapse during deep dives.

↘ Deep dives require partial collapse of the rib cage, made possible by numerous floating ribs and flexible rib joints.

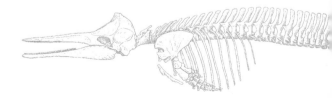

THE DECOMPRESSION PROBLEM

Another potential pressure-related problem is decompression sickness (DCS, also known as "the bends"), a well-known hazard for scuba divers who stay down too long or ascend too quickly (see Chapter 12, page 145). At depth, higher pressures allow more gas to dissolve in the blood than at the surface. If a diver ascends too quickly, the decreasing pressure causes the gas to bubble out of solution within the blood vessels before it can be expelled at the lungs. The bubbles can block circulation, causing injury or death. The main culprit is nitrogen, an inert gas that makes up 78 percent of air and is abundant in the blood.

DCS is a significant risk for scuba divers, who continually breathe compressed air at depth, but even free-diving whales can get DCS under the right conditions. While at depth, nitrogen in the lungs from a whale's initial breath at the surface can diffuse into the blood in larger amounts than it would at the surface. If the whale surfaces for a short interval and re-submerges for the next dive before it has outgassed all of the previously absorbed nitrogen, it will carry down a slightly elevated blood nitrogen content. If the whale does this over and over, it can gradually build up enough nitrogen in its blood to cause bubble formation.

~ Preventing bubble formation ~

Generally, cetaceans avoid DCS because their lungs collapse at depths of 164–328 ft (50–100 m), depending on the species. As the lung's alveoli air sacs collapse under pressure, air is pushed into the bronchiole air passages, which are held open with stiff cartilage rings. With no air in the alveoli, no gases are exchanged with the blood at high pressure, so blood nitrogen levels do not rise and DCS is avoided. A whale can still be vulnerable to DCS, however, if it dives repeatedly, with short intervals between dives, to depths slightly above where the lungs collapse.

SOUND AND HEARING IN WATER

Hearing is the most important sense for whales, effective across long distances and unhindered by dark or murky water. Sound travels farther and faster in water than in air. Low frequency (low pitch) sounds in particular can travel across whole ocean basins, and under the right conditions, whale calls can potentially travel thousands of miles. Whales are known to react to calls on the scale of tens of miles, but whether they actually share information over hundreds of miles or more is a fascinating and unanswered question.

FREQUENCY RANGES FOR HEARING

Most of what we know about hearing in whales comes from captive studies of smaller odontocetes. Humans can hear frequencies ranging from low-pitch sounds at 20 Hertz (Hz, wavelengths per second) to high sounds around 20,000 Hz (20 kHz). Mysticetes can generate and hear infrasonic sounds as low as 10 Hz, too low for humans to hear, despite being quite powerful. Their upper frequency range is unknown but unlikely to be much over 20 kHz. Odontocetes can hear across a much wider range of frequencies, from about 70 Hz to over 150 kHz, depending on the species. The emphasis on ultrasonic (greater than 20 kHz) sounds among odontocetes is important for echolocation (see page 64). Echolocation is an active sense in which sound is directed toward a target and the returning echo provides information about size, shape, and distance. Echolocation occurs in a handful of animal groups (sight-impaired humans have shown modest abilities), but truly advanced abilities are only found in toothed whales and bats.

→ This common dolphin (*Delphinus delphis*), which stranded alive with other pod members, is given a painless hearing test to assess its health prior to being cleared for release.

HEARING BY BONE CONDUCTION

Underwater hearing in cetaceans relies on sound conduction via bones and acoustic fats. The densities of water and the soft tissues of the body are similar, so sound waves in water pass right through soft tissues and reflect off the dense bones, causing the bones to vibrate. This is how humans hear under water as well, but our ears are not built for bone conduction, so many sounds are muffled and dull. Furthermore, the sound waves vibrate the entire skull, disguising whether the sound reaches the right or left ear first and eliminating our directional hearing. Cetaceans, especially odontocetes, have perfected hearing via bone conduction.

HEARING

Cetaceans receive sounds primarily via vibrations of the mandible, or jawbone. In odontocetes, the partially hollow jawbone is filled with specialized acoustic fats that provide exceptional conduction of sound directly to the left and right auditory bullae (ear bones), abutting the base of the jaw.

A defining character of cetaceans in evolutionary studies, the auditory bulla is comprised of two dense bones, the tympanic and periotic, which house the middle and inner ears. Suspended from the skull by ligaments and encased in foamy, oily tissue, the bulla is isolated from the disruption of additional acoustic vibrations from the skull. The ear drum has morphed into a calcified ligament that helps to transfer tympanic bone vibrations to the three tiny, middle ear bones within and, as in terrestrial mammals, these "ossicles" transfer vibrations to the cochlea, a fluid-filled, spiraling organ whose hair cells vibrate and fire, sending information about sound frequency and power to the brain.

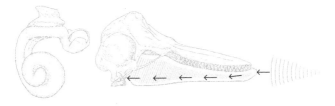

↑ Found within the periotic bone, the cochlea is the inner ear organ that interprets the frequency and power of sounds. Sensitive hair cells fire as they sense vibrations, sending information to the brain.

↑ Sounds are transmitted to the ear via specialized acoustic lipids in and around the jaw that help to conduct sound vibrations along the mandible, delivering them directly to the adjacent auditory bulla.

↙ The first hearing test of a baleen whale was recently conducted on a northern minke whale (*Balaenoptera acutorostrata*), revealing a wide frequency range, extending slightly into the ultrasonic range.

ECHOLOCATION STRUCTURES

E cholocation is an active sense—cetaceans must intentionally make sounds in order to generate echoes for interpretation. Although mysticetes may use a primitive version to interpret navigational echoes from the seafloor and major underwater features, only the odontocetes have evolved a precise underwater echolocation sense routinely used for fine-scale investigations and orientation.

ECHOLOCATION ABILITIES

Echolocation is often described as "seeing with sound." In controlled studies, captive dolphins can observe a complex shape under water using only echolocation and then distinguish it from a similarly complex shape using only vision, demonstrating both the fine-scale resolution of echolocation and the translation of information between echolocation and vision. Unlike vision, echolocation can provide precise measurements of distance, size, and even body composition, as the characteristics of echoes vary with the density and internal components of objects. Objects with large density differences, such as fish with air-filled swim bladders, are particularly good reflectors, while a large, colorful jellyfish with a density similar to water is much harder to detect.

SOUND GENERATION AND FOCUSING

Both observations and anatomy suggest that baleen whales generate sounds in their larynx like other mammals, but odontocetes produce both their echolocation clicks and communication (whistle) sounds from a pair of muscular "phonic lips" located between the skull and blowhole and associated with a system of nasal air sacs. Air moving into and between the sacs is forced through the phonic lips, causing them to vibrate and generate sound. Much of the sound passes forward into the adjacent melon, which is the large, fatty forehead of odontocetes. The melon fats are specialized, low-density lipids with excellent sound conduction properties, similar to those of the jaw. Like a hand lens with light, the melon serves as an acoustic lens, creating a focused beam of sound that proceeds from the front of the

THE UNIQUE SPERM WHALE DESIGN

Sperm Whales (*Physeter macrocephalus*) represent an intriguing exception to the general odontocete design. Their enormous head is dominated by two large features: the tapered, barrel-shaped spermaceti organ, encased within cable-like, fibrous strands and filled with over 500 gallons (about 2,000 liters) of liquid spermaceti oil, and the underlying "junk," a layer of oily tissue of lesser value to early whalers in comparison to the more accessible liquid spermaceti oil (see Chapter 11, page 140).

Winding between and around these features, one of the nasal passages leads to a blowhole at the front tip of the head and another leads to the phonic lips (A) and associated air sacs that generate sound at the anterior tip of the spermaceti organ (B). The sound is projected back through the spermaceti organ toward air sacs along the nearly vertical wall of the skull, where it is reflected forward again through the junk (C), which serves as the focusing melon for sperm whales. At peak power, the resulting beam of sound is possibly the loudest sound generated by any animal.

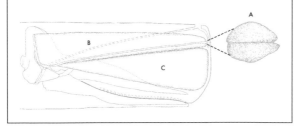

melon. Any generated sounds that are directed backward will be reflected forward again through the melon by the odontocete skull, whose parabolic shape was evolutionarily crafted for this purpose. Thus, the majority of the sound energy is directed forward into a highly directional beam in search of a target. When the echoes return, the sounds are received via the lower jaw, as previously described.

ECHOLOCATION IN ACTION

Since the determination of distance to a target requires a distinct start time and return time for sound, echolocation sounds are very short-duration "clicks." Odontocetes send out another click only after the first echo returns, with a series of clicks forming a rapidly repeating click train or "buzz." The clicks span a broad range of frequencies. Lower frequencies travel long distances, but their long wavelengths will not reflect off objects smaller than the wavelength. Higher frequencies, with shorter wavelengths, reflect off smaller objects and provide higher resolution, but they dissipate quickly, so have a limited range. Thus, a dolphin may initially search using clicks with a peak frequency (greatest power) around 7 kHz, capable of detecting an 8-in (20-cm) fish hundreds of yards away, but as the dolphin gets closer, peak frequencies will increase to provide greater detail, while the click repetition rate gets faster since each echo returns more quickly. At close range, hundreds of echolocation clicks (up to 150 kHz) are generated each second.

↓ Echoes return from sounds generated in the phonic lips (A) and focused by the melon (B). A 150 kHz frequency provides a resolution of less than 1/2 in (1 cm).

→ A striped dolphin (*Stenella coeruleoalba*) relies on echolocation to identify school size, movement patterns, and individual fish as it approaches a school.

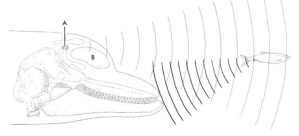

VISION

The eye lens commonly gets all the credit for focusing, but two-thirds of the focusing in human vision occurs when light bends (refracts) as it passes from air into the denser cornea and fluid-filled eyeball. The elliptical shape of our convex lens is designed to complete only the final third of focusing, delivering a crisp image to the photoreceptors lining the retina.

Ciliary muscles within the eye slightly deform the shape of the lens for fine-scale focusing of close or distant objects. Under water, however, the tissues and fluids of the eye are similar in density to the water, so the bending of light at the cornea is largely eliminated. Our elliptical lens is simply not strong enough to do all of the focusing by itself, so the focusing point under water ends up well beyond the back of the eye, resulting in a blurry image.

CETACEAN FOCUSING UNDER WATER

Physics works the same for whales—refraction by the cornea is negligible under water. However, whales (and fish) have solved the underwater focusing problem by having a spherical lens. The spherical shape bends light more strongly, focusing the image on the retina all by itself. Cetaceans lack ciliary muscles (a spherical lens is not easily deformed), so any fine-scale focusing that may occur is thought to be caused by muscles shifting the position of the eyeball within the orbit, altering the intra-occular pressure and moving the lens slightly forward or back.

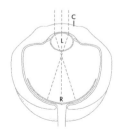

← Cross-section of a cetacean eye, showing: (C) the slightly flattened cornea, (L) the round lens, and (R) light projecting on to the retina. The closing of the pupil results in pinhole openings that assist with in-air focusing.

Pupil stages

open

closing

pinhole

LOW LIGHT AND COLOR VISION

Nearly all photoreceptors in the whale retina are rods, specialized for low light levels in deep or murky waters. When fully dilated under water, the extra-large pupil collects as much light as possible and a reflective layer behind the retina called the tapetum lucidum reflects light back through the photoreceptors a second time, acting as an efficient light multiplier. In addition to rods, 1–2 percent of cetacean photoreceptors are cone cells, required for color vision. However, the types of cones present are calibrated for blue or green wavelengths, the most abundant wavelengths in water, and this limited variety does not span enough wavelengths to allow a comparative perception of colors. Behavioral studies in aquaria also indicate that dolphins do not distinguish colors.

CETACEAN FOCUSING IN AIR

Amazingly, cetaceans also have good vision in air, even though their strong spherical lens should combine with corneal refraction in air to over-bend incoming light. The mechanism by which they avoid this problem is not fully understood and likely has multiple components. Many cetaceans have a partially flattened cornea, minimizing the corneal refraction in air for perpendicular, incoming light. The cetacean pupil also constricts unconventionally, forming a U-shaped slit that closes almost completely in bright, aerial conditions, leaving two pinhole openings at either end of the U. Light passing through a tiny opening has a focusing effect, so a pinhole pupil or a narrow slit can potentially deliver focused light over a range of distances. The pinhole openings line up with regions of high nerve cell density in the retina, suggesting alternative specialized focal regions for underwater and aerial vision. Light passing through these pinhole areas may also be directed along peripheral regions of the lens, potentially refracting differently than when passing through the core.

OTHER SENSES

Though hearing, echolocation, and vision have dominated researchers' attention, whales also rely on other senses, including chemoreception (taste and smell), touch, and, at least in some species, an electromagnetic sense.

SMELL AND TASTE

The sense of smell is understudied in whales. It has been completely lost in odontocetes, but mysticetes still have olfactory receptors, along with an olfactory nerve and bulb (brain region). Given their lack of echolocation abilities, a sense of smell may help mysticetes identify productive feeding waters, which often have a rich, organic, fishy odor.

The sense of taste is present in both toothed and baleen whales, with taste buds in localized areas on their tongues. Cetaceans have evolutionarily lost the primary genes for nearly all major taste categories (except "salty"), but in behavioral studies, trained dolphins have responded to each category—although their sensitivity is often very low. Despite these results, dolphins routinely press their rostrums to genital areas and swim through excretions of other dolphins, leading to speculation that they may identify excreted hormones and other signaling molecules based on taste.

TOUCH

The tactile sense is most developed in two regions of the cetacean body: the area surrounding the blowhole, which facilitates the timing of breathing as the animal clears the surface, and the sensory hairs (vibrissae) and empty hair follicles of the rostrum and chin. Humpback whales (*Megaptera novaeangliae*) have an array of fist-sized bumps (tubercules) on their rostrum and chin, each with a single, stiff vibrissa hypothesized to identify vibrations from prey aggregations in the water. Most dolphins have vibrissae along their rostrum that fall out before or soon after birth, although Amazon river dolphins (*Inia geoffrensis*) retain them as adults. The empty follicles remain well enervated for life, and dolphins will sometimes investigate objects by rubbing their rostrums against them.

ELECTRORECEPTION AND BIOMAGNETISM

Interestingly, the empty rostral hair follicles of bottlenose dolphins (*Tursiops* species) and Guiana dolphins (*Sotalia guianensis*) have recently been shown to develop into electroreception organs in adults, capable of detecting weak electric fields like those produced by the muscular ventilation of fish gills. Although not quite as sensitive as sharks to electric fields, this ability is another tool for locating hidden bottom-dwelling prey at close range. It also suggests a potential mechanism for a biomagnetic navigational sense, long hinted at by studies correlating live whale stranding patterns with the orientation of magnetic field lines and other geomagnetic anomalies.

↓ Small hairs on the rostrum and chin of this southern right whale (*Eubalaena australis*) provide a tactile sense that has been hypothesized to aid in sensing feeding currents.

SKIMMERS AND GULPERS

Depending on the species, mysticetes have 200–400 baleen plates on each side of their upper jaw. The size and structure of the baleen, and the entire body plan, is closely linked to the feeding technique they use.

SKIMMERS

The balaenids—right whales and bowhead whales (*Balaena mysticetus*)—are skimmers. These whales, which reach lengths of around 45–60 ft (14–18 m), hold their mouths wide open as they swim through dense zooplankton patches of mainly copepods (crustaceans less than half the size of a grain of rice), or somewhat larger, shrimp-like krill. Water enters at the front and filters out the sides through their 8–10-ft (2.5–3-m) long baleen. Emerging from the gums and made of keratin, baleen is akin to an enormous fingernail hanging

from the roof of the mouth. Each piece grows continuously and wears down to a tapered point. The inner edge frays into fine hairs and the plates stacked side by side essentially form a nice head of hair that serves as a fine filter.

GULPERS

The rorqual whales are sports cars by comparison. Streamlined and moving swiftly through the water, they are gulpers, only opening their mouths when they lunge at a school of fish or krill. Whales basically use the current to swim their giant mouths open during lunge-feeding, opening them a full 90 degrees and expanding their pleated throats as they take in a huge volume of water. The enormous water pressure would shatter a normal jaw, but they mitigate the stress with flexible connections at the jaw hinge and at the tip of the jaw where the left and right mandible meet. The water is not released until they close their mouths most of the way and filter it through short baleen as it exits. The short baleen facilitates a flatter, more streamlined head. A blue whale (*Balaenoptera musculus*), for example, can be over twice the length of a right whale, but its baleen is less than half as long. In some rorquals that target fish instead of tiny plankton, the baleen forms coarse bristles rather than hairs.

THE UNIQUE GRAY WHALE

Gray whales (*Eschrichtius robustus*) are unusual, being the only baleen whales that are primarily bottom-feeders. They use suction to scoop up sediments while swimming on their sides, leaving furrows in the bottom of the seafloor and filtering out amphipods, plankton-sized crustaceans that live in and on the bottom.

← Southern right whale (*Eubalaena australis*) skim-feeding at the surface. The open mouth creates enormous drag—low speeds only capture slow-moving zooplankton.

PREDATION IN THE DEEP

Although deep-diving cetaceans consume many species of fish, their dominant prey items, by far, are deep-sea squid and octopus living in the dimly lit twilight zone (650–3,300 ft/200–1,000 m) and sunless waters beneath. Using echolocation extensively, sperm whales (*Physeter macrocephalus*) and beaked whales actively pursue individual prey at depth and target schools and spawning aggregations of squid when possible. Whales are estimated to eat more than 100 million tons of squid per year, and based on squid beaks in their stomachs, they consume over 50 species, ranging in size from a few inches to nearly their own length. The largest recorded giant squid was 43 ft (13 m) long, including two long tentacles extending about 16 ft (5 m) from the mantle (body), head, and arms. Colossal squids are slightly shorter, but their mantle is almost twice the size, making them the most massive invertebrate (greater than half a ton). Encounters with these two species are not rare—sperm whales eat millions of them each year.

↓ The sparse teeth of sperm whales are sufficient to secure squid. Beaked whales, with two expandable throat grooves, use suction feeding instead.

↓ Sperm whales often bear scars around their heads from the chitinous hooks and serrated suckers of giant and colossal squid tentacles.

→ The battle between sperm whales and giant squid, complete with flailing arms and tentacles, is one of our most enduring and romanticized visions of life in the deep sea—an amazing fact since no human has ever observed this event.

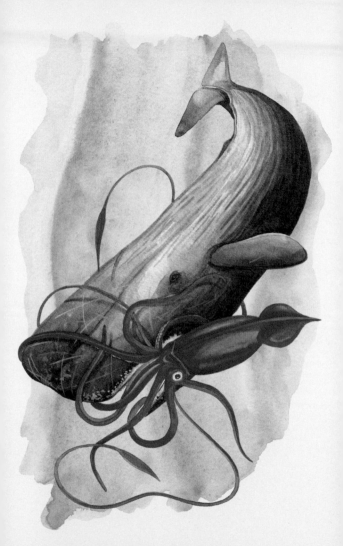

COOPERATIVE AND SOLITARY FORAGING

Most odontocete species increase their feeding efficiency through cooperative foraging, or coordinated group hunting behaviors. Strand-feeding Tamanend's bottlenose dolphins (*Tursiops erebennus*) in South Carolina, in the United States, are an interesting example. Residents of coastal salt marshes, they rush the muddy creek banks at low tide in groups of two to seven tightly packed dolphins, causing a bow wave that washes fish up onto the bank. The dolphins' momentum strands them on the bank temporarily, allowing them to pick individual fish off the mud before wriggling back into the water. Strand-feeders always strand on their right side, a lateralized preference that is common among many foraging cetaceans and possibly related to differences in visual specializations of the right versus left eyes.

BAIT BALL FEEDING

In a fascinating study in Hawai'i, a multi-beam sonar was used to investigate the behavior of foraging spinner dolphins (*Stenella longirostris*) feeding on fish schools in the open water column. Groups of 16–28 dolphins would encircle the fish, distributing themselves in equidistant pairs as they swam continuously around the school. Once the school contracted to a dense bait ball, two pairs of dolphins would

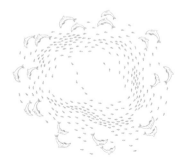

← A group of spinner dolphins exhibit intricately coordinated teamwork when driving a school of fish into a dense bait ball, sharing equally in both effort and reward.

enter the edge of the ball from opposite sides of the circle to feed for about ten seconds before returning to their position circling the school, at which point the next pair of dolphins immediately behind them would enter the bait ball to feed. The group systematically cycled through all dolphin pairs several times, providing multiple opportunities for each pair to feed in a band of fish with a density 60 times greater than the original school. One wonders how this astoundingly well-choreographed behavior was coordinated, as well as the consequences from the group for any individuals who didn't perfectly follow the script.

CREATIVE FORAGERS

Whether cooperative or solitary, the list of creative foraging techniques by whales is endless. Researchers have been equally creative describing them, with terms like kerplunking (a tail slap with follow-through, producing a loud "kerplunk" to startle fish), cratering (dolphins corkscrewing into bottom sediments in search of prey), and fish whacking (smacking fish with the flukes).

Amazon river dolphins (*Inia geoffrensis*) swim through the flooded forest during the rainy season, using their flexible necks and long, thin rostrums to grab fish from among the tangled branches and vines. Beluga whales (*Delphinapterus leucas*) squirt powerful jets of water from their mouths, potentially displacing hidden prey when they forage along the bottom of the seabed. Many species have learned to steal fish from nets, traps, and fishhooks on commercial long lines. In Georgia, in the United States, where shrimp bottom trawls have escape panels to prevent sea turtle bycatch, dolphins have been filmed inserting half their body through these panels into an actively towed net to grab captured fish as they come down the chute.

HUMAN–DOLPHIN COOPERATION

Cooperative foraging relationships even develop between dolphins and humans. In Laguna, Brazil, some members of a resident population of bottlenose dolphins (*Tursiops* species) regularly herd fish toward fishers who wait in shallow water with cast nets. Using specific head movements or tail slaps, the dolphins signal to the fishers when and where to throw their nets. The dolphins have increased success feeding on the escaping, disoriented fish, while the fishers maximize their catch with fewer wasted casts.

BUBBLE NET FEEDING

M ost baleen whales are solitary foragers within loosely defined groups, grazing individually on vast aggregations of plankton or fish. Cooperative foraging has been observed in a few mysticetes, including circular swimming groups of southern right whales (*Eubalaena australis*) and Eden's whale (*Balaenoptera edeni edeni*), a subspecies of Bryde's whale, but it is most notable among specific groups of humpback whales (*Megaptera novaeangliae*) from various locations. The best-studied of these are teams of bubble net feeders in Alaskan waters, and although bubble nets are commonly used by solitary humpbacks, here it is a team sport. In this behavior, one individual swims in a circle beneath a school of herring, blowing a ring of bubbles that encircles and traps fish near the surface. The whales then lunge-feed up through the center of the school, bursting from the water with their mouths agape. Some team members are multi-year associates and may play specific roles, including blowing the bubbles or executing various loops, spirals, and positions within the group.

↑ A humpback whale's circular net of rising bubbles is an effective barrier against escape for schools of fish.

↑ Humpback whales using bubble net feeding target herring, but common microscopic zooplankton prey for mysticetes include krill, copepods, and amphipods.

↙ A team of humpback whales engages in bubble net feeding. Groups can range from two to twenty whales. With their unusually long flippers, humpbacks are the most agile of the baleen whales for such tight, coordinated maneuvers.

THE APEX PREDATOR

If we leave humans out of the discussion, orcas or killer whales (*Orcinus orca*) are arguably the apex predator on the planet, occupying the top of the food chain with no natural predators. As a species, their diet includes other so-called marine apex predators, such as great white sharks, dolphins, and even sperm whales (though not the large males). Terrestrial apex predators feed mainly on large herbivores—for example, lions on wildebeests—while orcas are at the top of longer food chains in which the typical herbivores are tiny zooplankton. Most food energy is lost to metabolism rather than growth, causing exponentially increased costs for higher levels in the chain. Since orcas feed about two steps higher than lions, the production of 1 lb (450 g) of orca requires about 100 times more plant and algal production at the base of the food chain than the production of one pound of lion.

ORCA SPECIALIZATIONS

Orcas are called killer whales because they kill other whales, not because they kill humans. There are no confirmed accounts of orcas killing humans in the wild. Different orca pods (groups) have preferred prey and foraging techniques.

~ Fish-eaters ~

Some pods specialize on schools of 12-in (30-cm) long herring, forming tight bait balls like spinner dolphins and stunning the fish with powerful smacks from their flukes. Other pods specialize on large salmon (18–38 in/46–97 cm), capturing fish individually, although they search for and share their prey as a group. Pods that specialize on sharks have worn-down teeth from the sharks' sandpaper-like skin. For large sharks, orcas eat only the large liver, full of calorie-rich oils.

~ Mammal eaters ~

Some pods eat primarily marine mammals, including seals, sea lions, dolphins, porpoises, other odontocetes, and baleen whales. Pods will even kill adult blue whales, chasing them for hours and exhausting them by biting at the blowhole, which makes it difficult to surface and breathe. They eat mainly the tongue of the whale, which may seem wasteful, but a blue whale's tongue can be as large as an orca. In Antarctic waters, groups of orcas will swim side by side to form a bow wave to wash resting seals off floating pack ice and into the water, while in Patagonia, solitary beach-hunting orcas will pluck sea lions from the edge of the surf zone in an attack reminiscent of strand-feeding dolphins.

↑ Swimming close together on their sides, a group of orcas creates a bow wave that will wash across the pack ice, sweeping the helpless crabeater seal into the water for easy capture.

ECOLOGICAL ROLE OF CETACEANS

Cetaceans command a significant portion of ecosystem resources due to their high position on the food chain and their high metabolism as warm-blooded mammals. Estimates of the proportion of total annual photosynthetic production from various oceanic regions required to support the cetacean food chain range from 26 percent in the North Pacific for just great whales, to 20.4 percent in the Gulf of Maine for all cetaceans, to 3–7 percent in a South Carolina salt marsh estuary for just bottlenose dolphins. Annual fish predation by whales exceeds the human fishery harvest in these oceanic regions. Recent studies using improved technology to estimate consumption rates and prey abundance suggest that mysticetes eat about twice as much as previously believed, indicating the oceanic percentages above are underestimates.

IMPACTS ON PREY POPULATION

Despite their large energy demand, the impact of cetaceans on the structure of marine communities is not always obvious. Many odontocetes are generalist predators, and when they are absent, other marine predators may fulfill a similar role. In a few cases, however, changing patterns by whales have been shown to have dramatic impacts on specific prey species. The equivalent of just 3.7 orcas switching to feeding on sea otters was estimated to be the likely cause of a dramatic decline of roughly 40,000 sea otters in western Alaska across much of the 1990s. In a more recent example, just two male orcas caused a dramatic shift in the distribution of great white sharks

← Sampling a volume of water similar to the volume of its own body, with each gulp, a typical blue whale (*Balaenoptera musculus*) in the north Pacific can consume 17½ tons of food per day.

in parts of South Africa. The orcas arrived in 2015 and began killing great white sharks for their livers. The shark population soon shifted its distribution away from the area, impacting the local ecosystem and devastating a thriving tourism business built around great white shark tours and cage dives.

INDIRECT IMPACTS

Whales have many indirect impacts on their ecosystems, as well. The survival of many seabirds depends on fish brought to the surface by cetaceans feeding on bait balls, and in shallow waters, wading birds like great egrets can meet their entire metabolic needs by specializing on leftover fish from strand-feeding dolphins. When whales feed at depth and then defecate at the surface, they bring valuable nutrients to the sunlit surface, serving as fertilizers to stimulate phytoplankton growth. Some models suggest that the much larger pre-whaling populations of great whales were only sustainable due to a higher level of global ocean productivity created by their increased nutrient recycling.

Even dead whales play a role, as periodic "whalefalls" bring valuable and long-lasting food resources to the cold, sparse, deep-sea bottom, supporting whole communities and providing an essential stepping stone for deep-sea hydrothermal vent species whose planktonic larvae can settle, mature, and distribute more larvae of their own before the resource is used up.

SIGNS OF INTELLIGENCE

Humans have never been smart enough to agree on a universally accepted definition of intelligence, but we are very good at giving examples. Solving novel problems based on past experience is a clear sign of intelligence, as are coordinated or cooperative behaviors with other individuals, especially among social odontocete predators. Examples of both of these were on full display in many of the foraging behaviors from the previous chapter. Sponge-carrying dolphins demonstrate tool use, considered a sign of intelligence, and dolphins and whales in captivity can be trained to perform complex tasks, including the ability to make autonomous decisions when completing activities in open-water situations, such as search and retrieval and mine detection operations. They are efficient audible and visual learners, and learn quickly through imitation and observation.

COMMUNICATION AS A SIGN
OF INTELLIGENCE

Communication, however, is perhaps the indicator of intelligence that has received the most attention in whales. Communication is the successful exchange of information between individuals. To be successful, communication requires comprehension by the receiver, and many (but not all) researchers also add intent by the sender. In animal behavior studies, comprehension can be verified by an observable response, and although intent can be more challenging to confirm, it is generally accepted if the communication has an apparent benefit or enhances evolutionary fitness.

~ Vocal and non-vocal communication ~

Cetaceans are quite vocal, producing an array of whistles, clicks, squawks, moans, grunts, bangs, and more. Not all communication is vocal, however. Swimming postures and speeds can visually signal playfulness, aggression, or mating intentions, and subtle tactile contact while swimming is believed to provide social cues in dolphins. Non-vocal auditory communication is also important, such as a mother's sharp tail slap on top of the water to bring her calf back in line, or potentially the loud splash from a breaching whale that may be heard over a long distance. This chapter, however, will focus on several well-studied areas of vocal communication in whales, including the song of humpback whales (*Megaptera novaeangliae*), the dialects of orcas (*Orcinus orca*), and the signature whistles of dolphins. These and other vocalizations, in combination with coordinated group behaviors and problem solving, are why cetaceans are widely recognized as highly intelligent animals.

← Some bottlenose dolphins (*Tursiops* sp.) in Shark Bay, Australia, collect and wear marine sponges on their rostrums, presumably for protection while foraging along the bottom.

DOLPHIN SIGNATURE WHISTLES

Every bottlenose dolphin (*Tursiops* species) produces a unique whistle pattern and this is its most common vocalization. Lasting from less than one to several seconds, this signature whistle serves as an individual identifier, much like a name, and recognition of individuals based on their whistle has been demonstrated in wild dolphins. Developed within the first few months of life, one to two-thirds of all vocalizations produced by wild dolphins are signature whistles. Dolphins sometimes copy another dolphin's signature whistle and use it to address the other individual. This is primarily seen between mothers and their calves and between one or two long-term male companions. It may seem odd to go around repeating one's own name over and over, but membership in dolphin groups can change from hour to hour, so simply announcing one's presence may strengthen group cohesion when out of visual contact, and there is also value in providing a specific location in three-dimensional space.

↓ Examples of two different signature whistles. Lines moving up and down represent whistle pitch rising and falling with time

(top: ~0.5s, bottom: ~1.5s). The stacked lines represent harmonics of peak frequencies, a structural component that is visually obvious but difficult to discern by ear for humans. When a dolphin is in isolation, about 90 percent of its vocalizations are signature whistles.

→ Although best studied in bottlenose dolphins, evidence for signature whistles has been found in the narwhal (*Monodon monoceros*) and at least ten dolphin species, including this Atlantic spotted dolphin (*Stenella frontalis*) mother–calf pair. Signature calls may also occur in beluga whales (*Delphinapterus leucas*).

DIALECTS

Human dialects, such as a British or American accent, result from many generations of relative isolation from other groups. Geographically isolated populations of some marine mammals and birds have also developed distinctive variations, or dialects, for shared calls. Dialects in killer whales and sperm whales (*Physeter macrocephalus*), however, are unique in that they have developed between different groups that frequently interact.

SPERM WHALE VOCAL CLANS

Sperm whales communicate using short (less than three seconds), patterned sequences of echolocation-like clicks called codas. Adult males are solitary or in small groups for most of the year, so the fundamental social units for sperm whales are long-term, largely matrilineal groups of females (members are related via the female line) and their offspring. Each unit has about a dozen reproductive females, while males leave the group at around six to nine years old.

A social unit routinely uses 20–30 codas in their vocalizations, but not every unit uses the same types of codas. For example, 7 different

---

INSIGHTS INTO SOCIAL BONDS

The complexity of these communities and dialects implies specific use and meanings for the various calls, but their meaning remains unknown. The fact that different dialects have developed between groups of whales that regularly interact, however, is quite remarkable and provides great insight into the strong social bonds of each individual pod. Dialects likely enhance the identification and cohesion of the groups and may help to avoid inbreeding in orcas, since mating does not occur within pods. In orcas, in particular, dialects highlight each pod's reliance on internal communication and traditions in shaping all aspects of their behavior.

dialects of coda types exist among the approximately 300,000 sperm whales in the Pacific Ocean, each with stereotyped differences in the rhythmic patterns of clicks. Individual female social units will sometimes temporarily group up with other units, but they will only interact with units that share their coda type. Thus, the different dialects define different vocal clans, such that the regional population is vocally (but not geographically) divided into multiple vocal clans, each consisting of potentially hundreds of matrilineal social units. It is not clear why these clans arose or why different clans do not interact, but continuity is ensured as calves learn their clan's codas from their social unit during their first few years of life.

ORCA POD DIALECTS

Orca dialects are even more unique, and although known from various regions, they have been best studied in the primarily salmon-eating resident killer whales of the northeast Pacific. These life-long pods of whales, ranging from a few to several dozen members, are composed of one to several matrilines consisting of an older female, all her surviving male and female offspring, and any further descendants of her daughters, granddaughters, and so on. Up to five generations of family can be represented. Pods have a repertoire of 7–17 short calls (1–3 seconds). These pulsed calls are shared between pods, but each pod has its own distinct dialect with subtle variations. Some pods share similarities in their repertoires and are grouped into vocal clans. Genetic studies have shown that these clans are closely related, several generations removed from a past when they were all one pod, before it grew and splintered into multiple pods.

→ Spectrograms for two different dialects for a call shared by two different orca pods. One has more wavering frequencies at the start.

HUMPBACK WHALE SONGS

S everal species of baleen whales sing songs, including bowhead whales (*Balaena mysticetus*), blue whales (*Balaenoptera musculus*), and fin and minke whales (*Balaenoptera* species), but the most complex and well-studied songs belong to the humpback whale. Songs are stereotyped vocalizations that are repeated. Singing is done by male humpbacks, primarily in their tropical breeding and calving grounds. The exact purpose is unknown, but theories include: to establish territorial spacing between males, to advertise male qualities such as size to potential mates, or simply to attract females.

Humpback songs last 8–22 minutes and are composed of four to eight ordered themes. Each theme is composed of various sound units that may be ordered into specific phrases that are repeated multiple times, and song sessions may last up to 22 hours! Different breeding populations (such as in Hawai'i or the Caribbean) sing different songs, and within every population the songs evolve, replacing some themes or phrases each year and eventually replacing the entire song after two to eight years.

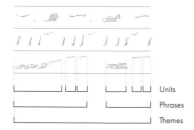

Units

Phrases

Themes

← A portion of a humpback whale song, showing the detailed structure of longer themes, made up of smaller phrases and units. The lines moving up and down represent the pitch of the song rising and falling with time (in this case, over several minutes).

← Male humpback whales assume a head-down position when singing. Males of other baleen whales also sing songs, including bowhead, blue, fin, and minke whales.

DO CETACEANS HAVE LANGUAGE?

Perhaps nothing fascinates people more about whales than the possibility that they might have their own complex language and that someday we might figure out how to talk to them. Virtually every animal is able to communicate at some level, whether it is a simple call of alarm or a chemical pheromone indicating it is time to mate, but "language," as the most complex form of communication, is truly rare.

WHAT IS LANGUAGE?

At its core, language requires two fundamental components: "words" (discrete signals—auditory, visual, or other—associated with a defined meaning) and syntax (the rules and grammar by which a set of words are combined in various orders and combinations to produce multiple meanings). Growing a vocabulary of words takes time, but syntax is even more challenging. A dog that knows the word "walk" is ill-equipped for the syntax and expanded vocabulary of "let's go for a walk today and tomorrow" versus "let's go for a walk tomorrow but not today."

LANGUAGE STUDIES IN CAPTIVITY

Bottlenose dolphins in captivity have been taught rudimentary languages using both acoustic and visual signals. They can amass vocabularies of over 100 words, including nouns (ball, water), verbs (fetch, swim), and adjectives (round, big, bigger), and can correctly interpret the syntax of novel strings of up to seven signals, such as the equivalent of "fetch the black ring and bring it to the white ball." Other animals have demonstrated similar or better language skills, including apes, African grey parrots, and sea lions, but while a two-year-old human follows up these early milestones with an explosion of accelerating language skills, the animal studies all plateau.

DO WHALES HAVE A LANGUAGE?

It is possible that the artificial languages constructed by human researchers simply don't play to dolphin's strengths. If so, is there evidence that any cetaceans use a language of their own in the wild, with a substantial vocabulary and syntax? Dolphins and orcas make additional whistles beside signature whistles and dialect calls, and beaked whales and some porpoises use pulsed or click-based communication sounds. Beluga whales are nicknamed "sea canaries" due to their frequent and diverse vocalizations, yet little is known about most of their sounds other than their acoustic structure.

ARE DIALECTS AND SONGS SIGNS OF LANGUAGE?

Both orca dialects and humpback whale songs have some components of language. The fact that a defined set of calls are retained throughout an orca population, with each pod applying their own dialects to the calls, suggests that these calls are important signals with widely recognized meanings (that is, potential words). There is no evidence that these calls are ordered according to any apparent syntax, and if they were, it seems odd that their vocabulary has not grown beyond about ten calls per pod.

Humpback whale song is particularly enticing, as the large number of multi-unit phrases and carefully ordered themes are reminiscent of words arranged in syntax. If this is true, however, the creative flexibility afforded by a vocabulary and syntax are never realized. Only one song is sung each year, so it appears the message never changes. In addition, if humpbacks do have advanced and flexible language skills, it is odd that only one sex participates. After many decades of searching, no firm evidence has been found for the use of language in wild cetacean populations, but researchers continue to investigate.

↘ Like most cetaceans, harbor porpoises (*Phocoena phocoena*) vocalize often, though their purpose or meaning remains unknown.

CULTURE IN WHALES

Cetaceans, along with species such as social primates and elephants, are some of the few animals described as having "culture." Culture in animals requires information and behavioral traditions that are broadly shared and passed down within a group through social learning. These traditions are persistent over generations, leading to long-established, behaviorally distinct subgroups within a species in terms of how they interact socially, communicate, forage, and distribute.

CULTURE IN THE MATRILINEAL WHALES

In cetaceans, cultural discussions focus mainly on the matrilineal whales: orcas, sperm whales, and the lesser-studied pilot whales (*Globicephala* species), whose social structure is known more from genetic analyses than behavioral studies. Vocal clans of sperm whales and vocally distinct orca pods identify groups that are also distinct in their foraging techniques, distribution, and general behaviors.

ORCA POD CULTURE AND ECOTYPES

The resident orca pods of the northeast Pacific are the best studied in the world. Their culture is remarkable in that all pod members are a multi-generational family, with both males and females remaining with their mothers for their entire lives and continuing to remain with the matriline if their mother passes away. Their behaviors and dialects are shaped by this internal focus on the family group, establishing a culture that lasts for generations.

The North Pacific, however, is also home to transient and offshore orcas, and their culture is quite different. Transient orcas eat mainly marine mammals. They are much less vocal than the fish-eating resident pods and their calls do not have pod-specific dialects. Although some individuals associate with their mothers for long periods, other males and females permanently separate from their matriline, facilitating a variety of distribution options, from roving pairs to large, temporary aggregations. In other parts of the world, the resident and transient labels don't work well and researchers have devised different categories.

~ Orca ecotypes ~

At the global scale, perhaps the best explanation is that killer whales have evolved at least ten unique "ecotypes" worldwide: distinct body size, shape, and coloration patterns representing populations (such as the North Pacific residents) that have behaviorally segregated from one another for a long enough period that even the evolution of their bodies reflects their long-term cultural foraging and behavioral traditions.

← Orca pods pass along from generation to generation their cultural traditions of foraging techniques, vocalization patterns, and seasonal movements.

SOCIAL STRUCTURE

Mysticetes typically do not live in large social groups like the odontocetes. They usually form loose associations of two to six whales, although larger groups of more than twenty whales have been observed cooperatively feeding when food is abundant. Mysticetes establish strong mother–calf bonds. This likely reflects their need to feed individually because small areas can't sustain the large daily food requirements of a group.

With the exception of bowhead (*Balaena mysticetus*), Bryde's (*Balaenoptera edeni*) and pygmy right (*Caperea marginata*) whales, most mysticete species migrate each year from summer feeding grounds in cool, high latitudes to winter calving and breeding grounds in warmer latitudes. Gray whales (*Eschrichtius robustus*) have the longest migrations of any mammal, with one tagged individual traveling 14,000 miles (22,530 km) from north of the Bering Strait to western Mexico. Humpback whales (*Megaptera novaeangliae*), which are abundant and more widely distributed, travel nearly as far, with separate breeding stocks that can overlap with other stocks on the feeding grounds.

← Bowhead whales socializing. These large groups of whales are most often seen when feeding, which is coordinated by sound.

↓ A mother–calf pair of gray whales, showing their dark gray coloration. Their bodies are covered with light-colored blotches and white patches of whale lice, especially on the head and tail. Calves stay with the mothers for approximately eight months.

MATING SYSTEMS

T he mating behavior of mysticetes is tied to a life cycle of summer feeding and winter breeding. Mating takes place in warm waters in winter. Mysticete females are larger than males, which helps them deal with the high energetic costs associated with rapid fetal growth and lactation. Males are not always competitive on the breeding grounds and in some studies show cooperation in mating. There is also evidence of strong bonds between older females, suggesting their value to whale societies.

BOWHEADS AND RIGHT WHALES

Male bowhead whales and right whales (*Eubalaena* species) have disproportionately large testes compared to their body size and larger testes produce more sperm. Testes size in right whales are the largest among mammals and can reach more than 2,000 lb (910 kg)! In order to keep the testes cool from the body's heat, they are equipped with a network of blood vessels that helps to control their temperature by bringing blood from the cold extremities.

Males do not engage in aggressive encounters, but instead employ sperm competition in which multiple males try to mate with a female in an attempt to "swamp" her with their genetic material to improve their chances of siring offspring and to dilute the sperm of competitors. Usually, the same males reproduce successfully each year, thereby reducing the gene pool and diversity, which adds to the vulnerability of the species.

HUMPBACK WHALES

A principal feature of the mating behavior of some mysticetes such as humpbacks is the performance by males of complex songs of extended duration to attract females on the breeding ground and to signal their status and readiness to mate. During this time, males often aggressively compete with one another for mating rights. The males try to impress the female in various ways, including launching themselves out of the water (known as breaching). The male that wins access—the "principal escort"—attends the female. Other males, termed "secondary escorts," attempt to displace the "principal escort." These challenges often involve displays (including singing) and physical contact between "escorts" to fend off new suitors and protect their female.

← Large head and mouth of a bowhead whale, showing the characteristic upturned lip and white patch on the lower jaw.

POLYGYNY IN SPERM WHALES

Sperm whales (*Physeter macrocephalus*) are the most sexually dimorphic of any cetacean, with the males being considerably larger than females. Males also possess big heads and large teeth, which are common traits among animals with polygynous mating systems, in which a male mates with more than one female in a breeding season.

Closely related adult females and their offspring occur in stable matrilineal family units known as "nursery schools." These consist of about a dozen individuals that range throughout tropical and subtropical latitudes year-round. Most females spend their lives in the same unit. Unlike killer whales (*Orcinus orca*), young males leave their families between four and twenty-one years of age and travel as nomads in loose associations, known as "bachelor schools," in higher latitudes. It is the large, solitary, physically mature males, however,

---

SPERM WHALES AND ELEPHANTS: PARALLEL SOCIETIES

The life history, growth, and social structure of sperm whales resembles that of the African elephant. The two societies evolved independently but in parallel with one another. The African elephant is characterized by large size, big brains, matriarchal social organization, and "roving" bands of adult males. The largest brain of any land mammal belongs to the African elephant, while the sperm whale's brain is a little larger than that of the blue whale (*Balaenoptera musculus*), the largest animal ever to have lived. Large brain size is linked to complex cognition and social organization. Like sperm whales, African elephants live in highly structured matriarchal groups from which males are expelled after reaching sexual maturity. Although male African elephants usually live alone, they sometimes form small groups with other males, similar to the "bachelor schools" of sperm whales.

---

that participate in most matings, searching out sexually receptive females as they move among family units. The "roving strategy" of adult males may be assisted by long-distance vocal communication between traveling males and female social groups.

During this period, males will sometimes fight one another, as evidenced by body scars made by the teeth of other males, but they do not defend harems for long periods, as the long calving intervals (five to seven years) and low number of potentially receptive females in each pod do not support a harem defense strategy. Though aggressive interactions occur among males, their relative infrequency does not seem to support the scale of sexual dimorphism seen in sperm whales. Rather, sexual selection by female preference for larger males appears to be a more likely driver for the evolution of dramatic sexual dimorphism, possibly related to an attraction to larger males because they can generate more powerful vocalizations.

FAMILY UNITS AND CLANS

As previously described, the matrilineal family units are linked to clans comprising hundreds to thousands of individuals linked by their vocal dialects and cultural traditions (see Chapter 7, pages 86 and 92). There is communal care of the young within family units, with females nursing calves that are not their own. "Babysitting" by mothers of young sperm whale calves continues for extended periods of time and they will defend themselves against predators by gathering in defensive formations. The genetics of populations of sperm whales subjected to intense commercial whaling shows evidence of reduced fertility, fragmented family units, and reduced body sizes.

↘ A sperm whale family unit, consisting of females and their female and juvenile male offspring.

DOLPHIN FISSION-FUSION SOCIETIES

In contrast to mysticetes, odontocetes live in diverse social groups. Most dolphins live in schools characterized by long-term associations among individuals. School size is variable, depending on species, location, and feeding habits. Some oceanic dolphins occur in schools of hundreds or thousands of individuals. One characteristic of dolphins is that they live in fission-fusion groupings in which individuals associate in small groups that change in composition. At any given time, individuals may associate in small groups or travel alone.

MALE BONDING

One of the longest-term studies of free-ranging common bottlenose dolphins (*Tursiops truncatus*) took place in the Sarasota Bay community, off the west coast of Florida. For more than 50 years this dolphin community of 100 individuals has been observed living in small schools of socially interacting individuals. These groups are organized according to age, sex, and reproductive condition. The mother–calf bond is strong and may persist for many years. Female groups are also relatively stable. Males form hierarchical alliances for mate acquisition. In early adolescence some males form coalitions with other males that may last for several years or even a lifetime. Bonded males may improve mating opportunities or ward off predators.

MATE GUARDING

Coalitions of male Indo-Pacific bottlenose dolphins (*Tursiops aduncus*) in Shark Bay, off the west coast of Australia, form lifelong alliances. These male dolphins perform synchronized movements and displays while singing in unison to attract females. Each male dolphin belongs to a small group of two to three individuals. They look for mates together and when they need back up, a second group of allies joins in. Sometimes these coalitions aggressively mate-guard and forcibly herd females, sometimes abducting them from other male coalitions.

BOTTLENOSE DOLPHINS AND CHIMPANZEES

Bottlenose dolphins and chimpanzees have similar fission-fusion societies. In both groups, the males form the strongest bonds. Both show male hierarchical alliances that are principally formed to gain access to females. These similarities are accompanied by the frequent use of aggression directed at rival males. The sociability of females varies more within bottlenose dolphin communities than within those of chimpanzees, and this may relate to differences in the distribution of food resources.

← Highly social
Indo-Pacific bottlenose
dolphins, displaying spots
on the belly and robust
body shape characteristic
of the species.

MATRILINEAL PODS FOR LIFE

Both orcas (*Orcinus orca*) and pilot (*Globicephala* sp.) whales are organized into matrilineal societies in which descent is traced through maternal lines. Both form stable family and social groups.

ORCAS

Unlike sperm whales, the matrilineal pods of killer whales include males and females of all ages, and for many killer whales, especially those from ecotypes similar to the northeastern Pacific residents, most individuals are born into a pod for life. Male orcas are 15–20 percent longer than females, twice their weight, and have much larger dorsal fins and pectoral flippers. Despite the large size of males, the undisputed leader of the pod is the oldest female (or females). In most strongly dimorphic species, adult males typically do not get along during breeding seasons, but there is no indication of aggressive interactions between adult males within the pod. This is because mating does not occur within the pod, thus avoiding inbreeding. Female preference is for larger, older males and they account for most of the matings. Older males have more tooth rake marks, suggesting the possibility of increased aggressive interactions. Like sperm whales, however, their strong sexual size dimorphism appears to be driven more by female preference than by traditional male-male aggression.

The largest number of matrilines are found in resident pods that inhabit coastal areas where their primary food source is fish, with a preference for salmon. While residents hunt individually, they will share food among their immediate family. Females were observed as being the most likely to share prey, especially with offspring. It has been suggested that this is the reason males have higher mortality rates once their mothers and grandmothers die.

Pods show specific dialects and shared calls between pods suggests common ancestry. The top level of structure in resident killer whale society is the community, which is made up of pods that associate with one another. Communities are defined based on association patterns rather than maternal genealogy. Orcas have evolved complex culture and they learn from one another.

The matriline is also the basic social unit in mammal-eating transient killer whales and shark-eating offshores, the latter named for their tendency to be found farther out to sea than resident pods. Unlike resident killer whales, the offspring in transient killer whale populations disperse from matrilines for extended periods or permanently. As a result, transient matrilines tend to be smaller than those of residents. Founding matrilines from offshore killer whales form resident killer whale populations.

SOUTHERN RESIDENT ORCAS

The southern residents are the smallest of the "resident populations," found mostly off British Columbia, Canada, and in Washington state and Oregon, in the United States. There are three pods of southern resident killer whales, called the J, K, and L pods. These pods have different distributions and different movements during the year, traveling to various areas to take advantage of the seasonality of salmon runs. The ongoing 20-year decline of the southern resident killer whales is due to decreased prey availability, the presence of organic pollutants, and disturbance from vessel presence and noise.

PILOT WHALES

Other members of the Delphinidae family, long- and short-finned pilot whales (*Globicephala melas* and *G. macrorhynchus*, respectively) live in pods of 20–90 individuals. They are highly social and have a similar social structure to killer whales, composed of individuals with close matrilineal associations. The pods are stable and pilot whales grow to maturity in their birth pod. Males move between family groups. Huge multi-pod gatherings provide opportunities for males to mate with females from other pods.

→ Pods of long-finned pilot whales regularly occur in groups of dozens or hundreds.

RIVER DOLPHIN SOCIETY

A part from mother–calf bonds, river dolphins do not form long-term associations. Group size rarely exceeds 10–15 individuals and these dolphins are most often seen singly or in pairs. Males are sexually dimorphic and remain separate from females. In general, larger groups occur in principal rivers and lakes, especially when resources are concentrated and abundant. Smaller groups occur in constricted river channels and tributaries, which contain fewer resources. Other than mother–calf pairs, associations between individuals are thought to be temporary.

AMAZON RIVER DOLPHIN DISPLAY

Adult male botos (*Inia geoffrensis*), freshwater dolphins living in the Amazon and Orinoco Rivers, perform ritualized displays waving branches, and floating vegetation. Coupled with the male's bright pink coloration the displays are thought to attract females, so female choice may be an important component of their mating system. Male botos, which are 16 percent longer than females, have more tooth rakes and serious injuries, suggesting that aggressive interactions between competing males are not uncommon.

↙ Male boto, or Amazon river dolphin, carrying floating vegetation in a mating display to attract females.

→ The Yangtze river dolphin, or baiji (*Lipotes vexillifer*) was declared functionally extinct in 2007 due to habitat degradation. It was characterized by having a dark back and lighter underside, a long, slender beak, very small eyes, low triangular dorsal fin, and broad, rounded flippers.

REPRODUCTIVE STRUCTURES

Male reproductive organs (that is, the testes and penis) in whales are tucked into cavities within the abdomen in a genital slit to help them maintain their hydrodynamic shape and reduce drag. In females the genital slit is located closer to the tail fluke. Females have two mammary slits, one on either side of the genital slit, from which milk is forcefully ejected into the offspring's mouth. Like other mammals, female whales have two ovaries, a uterus, and a vagina, which lie in the abdominal cavity. In odontocetes the ovaries are held in deep ovarian pouches, but in mysticetes they are more exposed.

A peculiarity of most female whales is that the corpus albicans, a scar on the ovary that is a remnant of ovulation, remains visible for the duration of the whale's life, providing a record of past ovulations. This makes it possible to estimate the reproductive history and the ages of individual whales.

↓ Left ovary of a Pacific white-sided dolphin (*Lagenorhynchus obliquidens*) showing corpora albicantia, the number of which indicates reproductive age.

↓ The external sexual differences of male and female whales. (A) umbilicus, (B) genital slit, (C) anal slit, (D) mammary slit, (E) mammary gland.

→ A male orca (*Orcinus orca*) approaching a female from below in readiness for mating. Orcas are the largest and best known of cetacean species. They are characterized by their black-and-white coloration pattern, large dorsal fin, and large, rounded flippers.

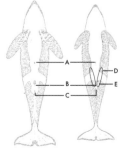

MATING, PREGNANCY, AND BIRTH

Whales mate underwater, belly to belly. Most whales have a multi-mate mating strategy, which is also known as a promiscuous mating system where males and females both have several mating partners in a single breeding season.

Most mysticetes have pregnancies lasting 10–13 months, while odontocete pregnancy lasts 7–17 months. Whales give birth to a single calf, which is born tail first. For mysticetes, birth occurs during marked periods of breeding that usually take place in winter and early spring in warm, shallow waters. For most odontocetes, births are seasonal with peaks during spring and summer.

Whale calves grow rapidly, but mysticete calves grow much faster than odontocete calves. In the largest animal on Earth, the blue whale (*Balaenoptera musculus*), fetal weight increases at a rate of 220 lb (100 kg) per day toward the end of pregnancy. The calf's weight increases at a rate of about 176 lb (80 kg) per day during nursing. During the seven-month period of dependence on cow's milk, the blue whale will have increased its weight by about 17 tons and increased its length from 23 to 56 ft (7 to 17 m). The faster growth of mysticetes is, in part, due to their migration pattern.

BLUE WHALES AND "HEAT RUNS"

Blue whales engage in a mating ritual known as a "heat run," which is also seen in humpback whales (*Megaptera novaeanglieae*). This behavior is instigated by females as a mating signal. It occurs when a female lures a group of males to give chase as they compete among one another by leaping, splashing, and fighting for mating rights. It is their way of demonstrating their energy and strength to both the female and their competitors. The conflicts can be quite intense, typically consisting of high-speed races with whales ramming into each other and pounding each other with their tail flukes.

NARWHAL TUSKING

The single, spiraled tusk of the male narwhal (*Monodon monoceros*) can be up to nearly 10 ft (3 m) in length, making it one of the most elaborate secondary sexual characteristics in mammals. "Tusking," a behavior in which males rub their tusks together (shown above), is thought to establish a dominance hierarchy among males that gives them an advantage when looking for a mate. Studies have shown that males with longer tusks are more successful in mating.

← Whales are born tail first, while the blowhole is the last to clear before the newborn calf heads to the surface for its first breath.

PARENTAL CARE

Many mysticetes are "capital breeders," which means that reproduction is based on stored energy reserves. After migrating from feeding grounds, individuals at the breeding grounds go through an elongated fasting period (lasting up to six months). Odontocetes pursue an aquatic nursing strategy in which mothers forage year-round.

Whale calves rely on their mother's milk, which is rich in fat and nutrients. Odontocetes have long nursing periods, ranging from 1–2 years and as much as 13 years for sperm whales (*Physeter macrocephalus*), while most mysticetes usually nurse for periods of less than a year. For several species of odontocetes, such as killer whales, pilot whales (*Globicephala* species), and sperm whales, at least some offspring remain with their mothers for life. The long nursing period of odontocetes has been related to social learning. A close mother–calf association is typical for many odontocete species. Indeed, outside of humans, no other mammal has such a prolonged and intensive maternal care investment.

↑ A female humpback whale and calf, showing their distinctive long flippers with serrated leading edge and prominent throat grooves.

↓ A close social bond between mother (below) and calf (above) is a characteristic feature of many odontocetes including common bottlenose dolphins

(*Tursiops truncatus*) illustrated here. This relationship is crucial to survival of the young because calves learn vocalizations during this time.

CALVING INTERVALS AND TIMING

Typically, whales do not calve annually but at intervals of two to three years or longer. Mysticetes generally give birth in the winter whereas odontocetes typically calve in the spring and summer.

MYSTICETES

Whale migrations are made by most mysticete species and some odontocetes, such as belugas (*Delphinapterus leucas*) and narwhals. The reproductive cycle of mysticetes is a minimum of two years, with reproductive events correlating with their annual migratory cycle. Most mating and calving take place on the wintering grounds at warmer, lower latitudes and calves are weaned when they arrive at the high latitude feeding grounds during their first summer. Females complete one migratory cycle when pregnant and the first half of the next cycle with a nursing calf.

Some species, such as gray whales (*Eschrichtius robustus*), migrate relatively close to shore, whereas others like humpback whales tend to migrate across ocean basins. Some species such as Bryde's whale (*Balaenoptera edeni*) appear to live largely in low latitudes. By contrast, some mysticetes, like bowhead whales (*Balaena mysticetus*), remain in the Arctic all year. It is possible that others, like the Omura's whale (*Balaenoptera omurai*), do not migrate.

~ Gray whale migration ~

Gray whales make the longest migrations of any mammal on Earth. Every year they swim more than 10,000 miles (16,000 km) in a round trip from their calving grounds in the lagoons of Mexico to their Arctic feeding grounds. Their migratory movements are relatively well known. For the spring northward migration (January to June), the first to head north from the breeding ground and return to summer feeding areas are the adult males, juveniles, and newly pregnant females. A pregnant female must put on enough fat to sustain her and her growing fetus through the coming year.

Mothers and their offspring stay in the lagoon for another month or two to gain strength before making the long journey. The southward migration (October to February) is led by pregnant females. Each year one-third to a half of adult females are pregnant. Late pregnant cows are among the first to return to the breeding and calving grounds, since early calving allows more time for a calf to grow. Soon to follow are the males and other adult females. Juveniles are the last to leave and some don't reach Mexico before turning and heading north.

While most gray whales migrate to Arctic feeding grounds, a group of several hundred individuals, known as the "Pacific Coast Feeding Group," spend their summer feeding off the shallower Oregon coast. Individuals in this group are smaller in body size, which may make feeding more efficient there compared to deeper Arctic waters.

ODONTOCETES

The reproductive cycle of odontocetes is more variable than that of baleen whales and typically lasts a minimum of three years. Unlike mysticetes, with their marked periods of breeding and feeding, odontocetes feed throughout the year. The cycle includes a one-year gestation and an approximately two-year lactation period. Occurring over several months are distinct events in the cycle: mating, calving, and weaning, when calves begin foraging but are still dependent on their mothers. Calving is timed to take advantage of food productivity in the spring and summer.

The movement patterns of belugas and narwhals, although often referred to as migrations, include persistent foraging, suggesting they may be better termed as long-distance movement strategies involving both migration and nomadism rather than classic migration.

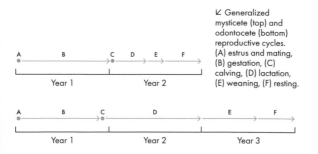

↙ Generalized mysticete (top) and odontocete (bottom) reproductive cycles. (A) estrus and mating, (B) gestation, (C) calving, (D) lactation, (E) weaning, (F) resting.

MATURITY AND LIFESPANS

Whales reach sexual maturity at different ages, depending on the species. The average age of sexual maturity in mysticetes varies between four years for humpback whales to up to approximately ten years for other balaenopterids—sei whales (*Balaenoptera borealis*), fin whales (*B. physalus*), and Bryde's whales—and up to 25 years for bowhead whales.

Among odontocetes, estimates of the age of sexual maturity range from three years for harbor porpoises (*Phocoena phocoena*) to fourteen years for killer whales and seven to ten years for sperm whales. Males and females of most species of whale become sexually mature within similar time frames.

In many species, males are larger than females, although among mysticetes females are generally five percent larger. The most pronounced sexual dimorphism is found in sperm whales, killer whales, bottlenose whales (*Hyperoodon* species), narwhals, belugas, and pilot whales (*Globicephala* species).

THE LONG LIVES OF BOWHEADS

Bowheads are the longest living whales. Based on the recovery of stone harpoon tips in harvested bowheads, as well as aging studies of eye lens protein, they are astonishingly ancient—up to 200 years old. The key to this longevity may lie in their genes. Scientists have recently mapped the genetic code of bowheads and the genes that confer long life could lead to drug therapies that fight aging and help humans live longer. Scientists have recently discovered that bowheads appear to hold the record for the longest pregnancies, at 23 months. Research suggests that bowheads and perhaps other mysticetes may be able to put a pregnancy on hold for up to nine months if conditions are not favorable.

LIFE SPAN

Whales are relatively long-lived and most live between 40 and 70 years. Generally, females live longer than males. The life span of a whale varies significantly based on the species, environment, habitat, and lifestyle. For example, whales in captivity are known to have shorter life spans. Similar to elephants, whales are k-selected species, which refers to the carrying capacity of the environment, and they share the following characteristics in addition to long life span: low reproductive rate, large body size, and low offspring mortality.

Among mysticetes, life spans range from 60 years in the common minke whales (*Balaenoptera acutorostrata*) and the Antarctic minke whale (*B. bonaerensis*) to up to 100 years for fin whales and 100–200 years for bowheads (see page 114). Among odontocetes, life spans range from around 20 years for harbor porpoises to 20–30 years for belugas to 70 years for sperm whales.

~ Estimating age ~

Age estimation methods are a critical tool for understanding and managing populations of whales. For effective management it is important to know how many individuals there are, their lifespan, when they mature and begin reproducing, and how fast they grow. A number of methods are used to estimate the age of a whale after it has died, including counting growth rings in the teeth, bones, baleen, or ear wax. For living whales, methods employed include photo identification, genetic markers, tagging, and analysis of fatty acids, some of which are associated with old age.

↘ Southern bottlenose whale (*Hyperoodon planifrons*) female (top) and male (bottom), illustrating sexual dimorphism exemplified by the larger size of the male.

MENOPAUSE

M ost animals continue to reproduce until they die. At a certain age, however, human females go through menopause and stop reproducing, but they continue to lead long, productive lives. Menopause is very rare among mammals and only three known species—killer whales, short-finned pilot whales (*Globicephalus macrorhynchus*), and humans—go through menopause. Short-finned pilot whale females live until their mid-50s but stop breeding in their mid-30s. Killer whale females stop breeding after roughly 48 years and can then live to the ripe old age of 90.

THE GRANDMOTHER EFFECT

Postmenopausal killer whales and pilot whales use their experience to help their families find food and rear young rather than birthing more offspring. The "mother effect" suggests that after a female has a certain number of children she puts them at a disadvantage by continuing to invest in the risks of childbirth, especially at an older age. Similarly, the "grandmother effect" suggests that older females can leave more surviving relatives by helping their children's children than by having more of their own.

Studies have shown that postmenopausal females put more energy into helping their sons, but not daughters or grandchildren. Evidence in support of this is that daughters are weaned at four to six years but sons are cared for into their teens. It makes sense for these mothers to help their sons because as the mothers grow older they are increasingly related to whales in other neighboring pods via their sons' calves. It is just the opposite for daughters whose closest kin are in the immediate pod. This also makes sense because her sons' offspring are elsewhere and so don't compete with her own for resources.

Another study showed that male orcas with a surviving postmenopausal mother have fewer signs of injury (that is, tooth marks on their skin resulting from fighting) than those whose mothers are dead or still reproducing. Research has also shown that there was no difference in scarring among young females regardless of the reproductive status of their mothers, meaning there was no special protection afforded to them.

← A pod of short-finned pilot whales that, along with killer whales, are organized into matriarchal societies comprised of females and their offspring.

ENTANGLEMENT AND BYCATCH

The incidental capture or bycatch of whales as the result of fishing activities is a global problem. Whales are subject to entanglement and drowning in nets, lines, and traps of commercial fishing gear. The sociality of some species of odontocetes, especially social delphinids, makes them more vulnerable to incidental capture. In the 1960s large numbers of pantropical spotted dolphins (*Stenella attenuata*) in the Eastern Tropical Pacific became trapped in nets intended for yellow fin tuna. This resulted in the mortality of an estimated six million dolphins. A redesign of the nets has greatly reduced dolphin mortality.

ACOUSTIC ALARMS

The use of acoustic alarms or pingers that warn animals about the presence of nets has reduced the number of entanglements. Pingers need to be used with caution since for some species they are not effective, while other species habituate to the alarm sounds. In addition, the effects of acoustic alarms can cover large distances and marine mammals might be deterred from certain habitats or migration routes.

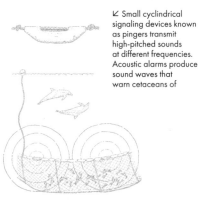

↙ Small cyclindrical signaling devices known as pingers transmit high-pitched sounds at different frequencies. Acoustic alarms produce sound waves that warn cetaceans of the presence of nets—in some situations, this has significantly reduced their mortality.

→ Spectacled porpoises (*Phocoena dioptrica*), with their distinctive two-tone color pattern, are often incidentally caught in fishing nets and drown. Entanglement in fishing gear and nets is also known as bycatch and a change in net design has significantly reduced porpoise mortality.

POLLUTANTS AND BIOMAGNIFICATION

One of the most widespread threats to cetaceans comes from pollutants. Researchers found that human activity such as the introduction of toxic chemicals into the environment has affected every square mile of ocean. Persistent organic pollutants (POPs) include hydrocarbons (benzene and methane); polychlorinated biphenols (PCBs), which are human-made chemicals previously used in the manufacture of various goods, including electrical equipment; pesticides (like DDT); and heavy metals (like lead and mercury).

Although banned in many countries, these pollutants continue to persist in the environment. For example, DDT has been banned in Western countries since the 1970s but is still detected in oceans and river systems due to its extreme stability. Recently, off the southern California coast, tens of thousands of barrels of toxic chemical waste containing original DDT was found from 50–70 years ago. These pollutants cause various reproductive abnormalities, nervous and digestive system problems, skeletal anomalies, and impaired immune function in marine mammals.

EFFECTS ON WHALES

Cetacean populations worldwide are affected by pollutants, including in remote regions like the Arctic and Antarctica, far from the sources of contamination, because these compounds are transported over long distances by atmospheric and ocean currents. Generally, cetaceans living in coastal areas are exposed to higher levels of pollution than those residing farther offshore and those occurring in industrialized areas usually have higher pollutant levels compared to those in less developed regions. Another factor in differential vulnerability is body size, with small species generally having higher levels of pollutants relative to their size.

As top predators in the ocean ecosystem, cetaceans are particularly susceptible to toxic chemicals, with their significant blubber layers acting as storage sites for these contaminants. The tendency for toxic compounds to accumulate and increase in concentration through

DEEPWATER HORIZON OIL SPILL

Catastrophic oil spills have detrimentally affected whales. The damage caused not only affects the animals directly but also their food supply through biomagnification. The largest oil spill in history occurred in 2010 when British Petroleum's offshore drilling unit Deepwater Horizon exploded, burned, and subsequently sank, spilling over a three-month period an estimated 3.19 million barrels (over 130 million gallons) into the Gulf of Mexico. In addition to the oil, nearly 2 million gallons (7.6 million liters) of toxic chemicals were dispersed into the Gulf. Assessments of marine mammals revealed increased mortalities (including strandings) of common bottlenose dolphins (*Tursiops truncatus*), spinner dolphins (*Stenella longirostris*), killer whales (*Orcinus orca*), sperm whales (*Physeter macrocephalus*), and melon-headed whales (*Pepenocephala electra*), as well as decreased reproductive success (only 20 percent successfully carry their offspring to full term) and increased health issues resulting from the toxic effect of oil exposure.

food webs is known as biomagnification. For example, toxins in plankton accumulate in small fish, which are in turn eaten by larger fish. Eventually, marine mammals higher up the food chain eat the toxic fish, further concentrating the toxins. Fat-soluble substances like POPs are particularly harmful since they accumulate in the tissues and are often transferred from mother to young during nursing.

THE CHANGING OCEAN

The increase of carbon dioxide and other greenhouse gases (so called because they trap heat) in the atmosphere is the main cause of global climate warming. Warmer water temperatures, rising sea levels, and increasing acidity in the oceans are inducing changes in the distribution of whales and marine food webs. These changes especially affect ice-associated species such as Antarctic orca, bowhead (*Balaena mysticetus*), and beluga (*Delphinapterus leucas*) whales.

Increases in temperature have resulted in the disappearance of krill in polar regions. Research has shown that krill levels have dropped 80 percent since the 1970s. Given the significance of krill to the diet of many baleen whales, its disappearance is cause for alarm.

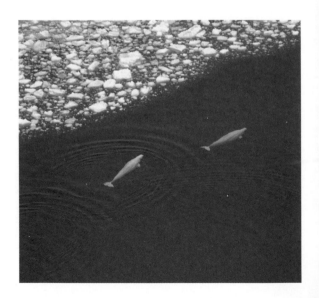

GRAY WHALES DELAY MIGRATION

Warming of the ocean has changed species ranges, including migration routes and the timing of migrations. For example, scientists have observed that gray whales (*Eschrichtius robustus*) delay their southward migration to Mexican breeding lagoons. As warmer temperatures melt the ocean's ice, other animals move into the gray whales' habitat and start feeding on their food, bottom-dwelling crustaceans. Crowded out by new competitors and with a dwindling food supply, gray whales start their migration later since they need to travel farther north and feed for longer to get their fill.

BELUGAS, BOWHEADS, AND MELTING ICE

Changes to ice coverage thickness affect the ability of belugas to migrate, following their typical routes, and more frequently to find food. Belugas must dive longer, deeper, and more often to feed and the resulting stress could reduce their reproductive abilities.

As ice declines, some bowhead populations are not migrating to the Bering Sea but are instead staying in the Canadian Beaufort Sea. This makes them vulnerable to ship strikes, underwater noise, and fishing gear entanglements.

HUMPBACK WHALES AND HABITAT LOSS

Humpback whale (*Megaptera novaeangliae*) breeding grounds are generally in tropical areas. Scientists predict that by the end of the 21st century, 35 percent of humpback tropical breeding grounds will experience a significant increase in temperature. For some humpbacks, their ranges will shift to areas close by, but for others living in isolated regions such as the Hawaiian Islands they may not find suitable habitats nearby.

← Beluga whales, with their distinctive all-white coloration, moving through Arctic ice floes. The whales migrate south in the fall as ice forms.

OCEAN NOISE

Human-generated or anthropogenic noise can detrimentally affect whales. Anthropogenic noise comes from many sources, including commercial shipping traffic, seismic and oil exploration, dredging, and military sonar. Noise pollution ranges from cargo vessels that emit sound frequencies of 20–100 Hz to military sonar at 30 kHz–500 kHz.

Whales have excellent hearing and can detect a wide range of sounds, often over long distances. The sound frequencies used by whales range from 10 Hz to 31 kHz. They use sound to communicate with each other, to avoid predators, and to find food. Anthropogenic sound can cause changes in whale behavior, ranging from minor orientation responses, to feeding less and producing fewer calls, to mother and calf separations. For example, female blue whales (*Balaenoptera musculus*) and their calves off the California coast traditionally travel nearshore, but they have shifted their distribution farther offshore in response to high volumes of ship traffic noise.

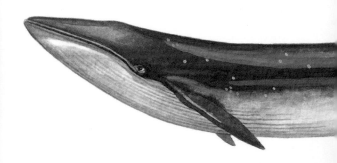

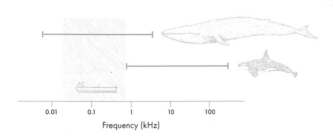

Frequency (kHz)

↑ The frequency of sounds produced by mysticetes and odontocetes compared to shipping noise (shaded rectangle) indicates the threat posed by human-generated sound.

↓ Shipping traffic noise interferes with the low-frequency vocalizations of Bryde's whale (*Balaenoptera edeni*), which disrupts the communication between whales and affects their ability to find food.

WHALING

Most species of whales have been hunted by humans at one time or another, although some have been hunted more intensively than others. The great whales, including sperm whales and baleen whales such as North Atlantic right whales (*Eubalaena glacialis*), bowhead, and gray whales, were among the first killed because they could be more easily hunted for their oil, meat, and blubber using hand-held harpoons.

Whalers from the Basque Country were the first to explore the North American coast. Initially, they hunted whales north of the Arctic Circle. The first commercial whale hunting in North America began in Labrador, where the first whaling station was found, dating from the 1500s.

WHALING IN AMERICA AND BEYOND

By the end of the 19th century "Yankee whaling," the oil industry of its day, which began in New England and the Atlantic coast, had spread to the Pacific and Indian oceans. Modern commercial whaling was initiated with the invention of cannon-fired, explosive head harpoons. Whaling in the United States hit its peak in the mid-1800s. In combination with the development of steam-powered catcher boats, whales could be taken in large numbers for the first time and large and faster swimming balaenopterids, such as blue whales and fin whales (*Balaenoptera physalus*), were taken.

The invention of factory ships in the early 20th century allowed harvested whales to be processed at sea, rather than depending on shore stations, and the numbers of whales taken increased dramatically. An estimated 2.9 million whales were killed during the 20th century. The great whales became scarcer, although the discovery of petroleum meant that whale oil was no longer needed. Global whaling peaked in the 1960s. Shortly thereafter, the larger whale populations had been decimated and efforts shifted to focus on the smaller common and Antarctic minke whales (*B. acutorostrata* and *B. bonaerensis*), which remained the main targets until the commercial harvesting of whales ceased during a moratorium enacted by the International Whaling Commission initiated in 1986.

Although the story of whaling is a story of over-exploitation, overall it is a conservation success since animals that were once in high demand across the world have been saved from extinction through international cooperation. Although populations of some species are still in recovery, other populations have increased to near pre-exploitation levels.

← Whaling in the 17th and mid-18th centuries targeted large whales that moved slowly near the coast and which could be easily hunted using harpoons.

MANAGEMENT AND CONSERVATION

Whales play an important role in the ocean's food web and in ensuring balance of the ocean's ecosystems. Whales face a variety of threats, from entanglement to collision with vessels, climate change, hunting, disease, pollution, and habitat loss. Conservation of cetaceans involves their protection and recovery by international and national policies and regulations. The International Union for the Conservation of Nature (IUCN) summarizes current information (updated annually) on the global conservation of plants and animals.

THE IUCN RED LIST

The IUCN maintains a Red List of Threatened Species and places species in categories of risk of extinction. Species are assessed according to geographic range, population size, and trends in population decline/increase, in addition to an analysis of extinction probability. According to the 2023 IUCN Red List, one in four whale species (20 percent of 92 species) are at great risk of extinction and are classified into the following categories: Extinct, Critically Endangered, Endangered, Vulnerable. The number of threatened cetaceans has increased from 15 percent in 1991 to 26 percent in 2023. Threatened odontocetes are principally located in Southeast Asia.

A study of drivers of extinction risk for marine mammals cited the vaquita or gulf porpoise (*Phocoena sinus*) as the most critically endangered cetacean, given its slow life history, small social groups, occupation of small geographic areas, exposure to entanglement risk, and estimated population of about ten individuals. With fewer than 100 breeding females and high entanglement rates, the Critically Endangered North Atlantic right whale was identified in the same study as being primarily threatened by ship strikes.

PROTECTING WHALES

Another important legislative regulation, the Marine Mammal Protection Act (MMPA) of 1972, established a framework for conservation measures, including a moratorium on the taking (that is, importing, harassing, hunting, capturing, collecting, or killing)

BANKS PENINSULA SANCTUARY

The area around Banks Peninsula in New Zealand is a sanctuary for the protection of the endangered Hector's dolphin (*Cephalorhynchus hectori*). Historically, Hector's dolphins (shown below), endemic to New Zealand, were victims of entanglements in gill nets. These are curtains of nets that hang in the water and result in the accidental trapping of marine mammals, including dolphins. Commercial gill net fishing was banned at Banks Peninsula Sanctuary, and subsequently, population numbers of Hector's dolphins have increased significantly.

of marine mammals in United States waters. Certain exemptions are allowed, notably that Indigenous people are permitted to hunt marine mammals, including some cetaceans—bowhead whales and gray whales—for subsistence purposes and for making and selling handicrafts made from whale products. Permits may also be granted for specific activities such as scientific research, public display, or taking animals to enhance their population and conservation status (for example, captive breeding).

When maintaining whales as sustainable functional elements of their ecosystems (one of the stated goals of the MMPA), it is important to have information on their population sizes, growth trends, and, for some species, formula-based quotas for allowable harvest. Important management tools for the monitoring of whales are the identification and addressing of threats, the promotion of research, community outreach, and sanctuaries, or protected areas for marine mammals.

GREEK MYTHOLOGY

Dolphin images appear frequently in ancient Greek art on frescos, sculptures, mosaics, and pottery, probably due to the role they played in Greek mythology and the fact that the Greeks were among the first great seafaring nations. It is through the writings of Greek poets that most of the myths about dolphins are known to us.

HOMER'S "HYMN TO APOLLO"

Many of the stories about dolphins link them closely with gods and describe them as having once been humans who were changed into dolphins by them. One of the earliest dolphin stories is found in Homer's "Hymn to Apollo," which describes how the Greek sun god Apollo, one of the favored sons of Zeus, founded the Oracle at Delphi after traveling all over Greece to find a suitable site and eventually choosing a cave at Mount Parnassus. The cave was guarded by the dragoness Python, whom Apollo slew with an arrow. Apollo in the guise of a dolphin set off to hijack a merchant ship. Terrified, the crew huddled below the deck while the dolphin Apollo directed the winds to blow the ship around the coast of Greece and into the harbor below Delphi. Then the sun god instructed his hostages to live in the new temple as priests and he carried them to this location on his back. After the founding of the temple at Delphi, many of Apollos' virtues were attributed to dolphins. As the god of shepherds and herdsmen associated with a pipe, flute, and lyre, he became the god of music and from this the dolphin of Greek mythology gained a reputation as a music-lover, in addition to being a messenger of the gods.

→ A red-figure vessel from ancient Greece, illustrating soldiers with shields riding dolphins, currently in the Metropolitan Museum of Art, New York.

DIONYSUS AND DOLPHINS

Several legends link Dionysus to dolphins. In one myth, Dionysus, conceived as the result of an incestuous relationship between Zeus and his daughter Persephone, is traveling in disguise on a pirate ship when the sailors onboard decide to sell him into slavery instead of delivering him safely home. Dionysus retaliates by driving the crew mad with hallucinations and they jump into the ocean. They are saved from drowning because they repent of their evil plan, at which point Dionysus relents and turns them into dolphins. This myth is often cited as the reason why dolphins were held in such high regard—killing them was equivalent to killing a person, and punishable by death.

BOTO ENCANTADO

Villagers in the Amazon relate stories about animals they encounter. The Boto encantado is an Amazon River dolphin (*Inia geoffrensis*), also known as the boto or pink dolphin, that lives in the murky waters of the Amazon River.

The people who live on the banks of the Amazon River have a long and complicated relationship with the species. While botos have a reputation for leading fishers to areas where fish are plentiful, botos also have a sinister side and may lure people to dangerous areas and situations. According to Brazilian folklore, botos transform at night from fish to handsome men, wearing a hat (to cover the blowhole), and visit drinking establishments and dance halls to seduce young women. If a woman becomes pregnant because she was assaulted by a community member or is forced into prostitution, a boto may be blamed to explain the situation. A fatherless child is referred to as the "child of the boto."

↓ Amazon River dolphins have long, narrow beaks, numerous pointed teeth, and very small eyes.

→ The Amazon River dolphin is distinguished by its coloration, which ranges from dark gray to bright pink (adult males), and its broad flippers and long, low dorsal fin.

INUIT MYTHOLOGY

The Inuits are a group of Indigenous people from the Arctic regions of Canada, Alaska, and Greenland. Whales play an important role in Inuit culture, both past and present.

NARWHALS

According to Inuit mythology the narwhal (*Monodon monoceros*) was once an evil woman with long hair that she had twisted and braided to represent a tusk. When the woman's blind son lashed her to a beluga whale (*Delphinapterus leucas*), she was drowned but transformed into a narwhal.

KILLER WHALES

Killer whales (*Orcinus orca*) are believed capable of changing into a wolf and after roaming about on land they may return to the ocean and become a killer whale again. According to another Inuit legend, a pack of wolves was floating on a patch of ice. The wolves, fearing death, asked the gods for help. The gods took pity on them and turned them into a pod of orcas. The Inuit viewed killer whales as wolves of the sea that were taught to eat seals and fish but not to harm humans.

BOWHEAD WHALES

An Inuit legend states that bowhead whales (*Balaena mysticetus*) were the Creator's favorite animals. They were offered to the people if needed for survival, but were not to be killed without a good reason. Bowheads provided the Inuit with valuable nutrients and sustenance given the scarce resources where they lived and the Inuit showed their respect by using every part of the whale.

SEDNA THE SEA GODDESS

One of the many versions of this myth originated among the Greenland Inuits. Sedna was a beautiful and clever woman. Hunters asked for her hand in marriage, but she always refused, saying that she didn't need a man to take care of her. This angered her father who demanded that she go to an island and stay there until she was willing to accept a prospective suitor. Sedna waited on the island and a handsome visitor asked for her hand in marriage. She accepted his proposal. One day, Sedna's father found out that his daughter had been tricked and had married an imposter. The father and family arrived to rescue her, but her husband transformed into a big eagle and created a massive storm. Sedna fell out of the boat and in an effort to save his family, her father cut off Sedna's fingers.

Sedna began living in the ocean. Her hair became tangled but because she had no fingers she could not comb it. Sedna's Inuit village was starving and they called upon a shaman for help. The shaman and spirits traveled to the bottom of the ocean and found a comb to brush Sedna's hair. She became a powerful spirit, the Sea Goddess. Ten Arctic animals swam out of her hair, one for each of her chopped-off fingers, including bearded seal, walrus, hooded seal, narwhal, reindeer, polar bear, ringed seal, musk oxen, harp seal, and many different birds. The Inuit believe that if they failed to live in harmony with nature and the animals, Sedna would punish them and they would be forced into starvation.

← Inuit myth in which an orca (*Orcinus orca*) is transformed into a wolf—orcas are viewed as the wolves of the ocean.

ST. BRENDAN AND THE WHALE

Featured in a Latin manuscript written in around 900 CE, an engraved map illustrates the mythical voyage of the 5th-century Irish monk St. Brendan, the patron saint of whales, who sailed from Ireland with other monks across the ocean 400 years before the Vikings and a thousand years before Columbus. The voyage took seven years. One of the most famous incidents on the voyage was when they came across a treeless "island" and decided to make camp for the night. The "island" began to move and the terrified monks fled. St. Brendan urged the monks not to be afraid since they were not on an "island" but had awakened the great fish Jasconius, which, in fact, was a large baleen whale (as evidenced by the throat grooves). St. Brendan held Easter mass on the back of this huge whale, and the monks built a fire atop the whale before it began to swim away. The print shown below is an important record of an early European illustration of a whale, demonstrating our longstanding interest in them.

↓ A seventeenth-century engraving of St. Brendan holding Mass on the back of a "sea serpent" that was likely a large baleen whale.

→ "Sea serpent" sightings may have been surface feeding whales such as this fin whale (*Balaenoptera physalus*). The whale is engaged in lunge feeding where the mouth opens and a throat pouch fills to consume fish and krill strained through its baleen.

DOLPHINS VERSUS SHARKS

Ah, the much-touted battle between sharks and dolphins . . . Traditional folklore claims that sharks fear dolphins and that dolphins will ram sharks with their rostrums and chase them out of the area to help protect their calves. In fact, if dolphins are aware of predatory sharks in the area and feel threatened by them, they can aggressively defend their pods in this way, especially those with young calves. The generally faster and more maneuverable dolphins are quite successful, especially when working cooperatively.

DO DOLPHINS PROTECT HUMANS FROM SHARKS?

Much folklore also exists regarding dolphins of various species assisting drowning humans and protecting human swimmers from sharks. Is there any truth to these stories? Actually, yes, there are a number of well-documented incidents, although one must wade through numerous sensationalized stories to find them. Examples include a dolphin pod that surrounded a sharkbite victim, shielding him from additional attacks as he worked his way to shore, and another pod that herded a group of swimmers into a tight ball when closely approached by a large shark, continuing the behavior for many minutes until the shark swam off. The motivation for these behaviors is unclear. Dolphins are social animals with a protective instinct for their own groups, but there is little chance they are confusing humans for fellow dolphins. Regardless, occasional incidents like these are more than enough to cement the "protective dolphin" concept into our folklore.

→ Whales and sharks often coexist with little drama, as seen with these short-finned pilot whales (*Globicephala macrorhynchus*) and this oceanic whitetip shark.

WHO'S WINNING?

Despite these examples, the perception that dolphins have the upper hand against sharks is incorrect. Simply put, most dolphins seldom eat sharks (mostly pups or juveniles), but plenty of sharks are trying to eat dolphins. Dolphin mouths are too small and their conical teeth are unable to cut off chunks of flesh, but the extendible jaws of sharks, full of large, cutting teeth (especially for species such as great white, tiger, and bull sharks) are quite capable of attacking dolphins. Shark bite scars are commonly observed in population level studies of dolphins. The highest rates are in bottlenose dolphins (*Tursiops* sp.) from the aptly named Shark Bay, Australia, where 74 percent of non-calf dolphins have at least one sharkbite scar, and calves have a 31 percent disappearance rate in their first year.

WHALING TALES

For more than a hundred years, beginning in the early 18th century, the heyday of American whaling centered on the ports of Nantucket and New Bedford on the Atlantic coast of the United States. These villages were filled with whalers, using wooden sailing and clipper ships to hunt sperm whales (*Physeter macrocephalus*).

The whalers were after the prized sperm whale oil, spermaceti, from the head cavity of sperm whales. Spermaceti oil, a liquid wax, was used to make candles. In addition to spermaceti, the head contains a spongy material impregnated with oil called "junk." The oil squeezed from the junk was of the highest quality and used to light lamps. The whalers' voyages in search of sperm whales, which lasted three to five years, took them around the world and provided rich stories of adventure based on encounters with sperm whales.

WHALING SHIPS

The real-life story of the whaling ship *Essex* was the inspiration for *Moby Dick*. On November 20, 1820, the *Essex* found a group of sperm whales in the South Pacific. The first mate, Owen Chase, harpooned a large male sperm whale, which thrashed about and smashed a hole in the 85-ft (26-m) boat. After several separate attacks, the *Essex* sank and 8 of the 21 sailors were rescued a few days later. Of the eight survivors, five went back to whaling.

Another account from 1851 tells the story of a ship called *Ann Alexander* being attacked and sunk by a sperm whale not far from where the *Essex* was struck. A wounded sperm whale struck the *Ann Alexander*, but the crew was rescued. Five months later, another ship, the *Rebecca Simms*, harpooned and captured a male sperm whale carrying a harpoon iron from the *Ann Alexander* and with splinters from the ship's timbers embedded in its head.

STAR OF THE EAST

A newspaper report features a sperm whale tall tale of the British vessel the *Star of the East*, which sighted a huge sperm whale in the Falkland Islands in 1891. Crew member James Bartley successfully harpooned the whale, which, when struck, dove several hundred feet. When it resurfaced, Bartley was missing and the crew reported movements in the belly of the whale. The next morning, the whale was cut open and out came James Bartley, still living after 15 hours in the belly of the whale. According to reports, Bartley's skin was deathly white, he had lost his hair, and he was blind. This account is purely fictitious, but the newspapers ran the story. The tale has continued to resurface over the years, with a story as recent as 1985 claiming that Bartley retired as a cobbler in Gloucester, England.

MOBY DICK

The best-known sperm whale tale is the 1851 novel *Moby Dick* (or *The Whale*) by American writer Herman Melville. The book is the sailor Ishmaels's narrative of the quest of Ahab, captain of the whaling ship *Pequod*, seeking vengeance on Moby Dick, a white sperm whale that had attacked Ahab on a previous voyage and bitten off his leg (shown below). The lone survivor of the ensuing encounter with Moby Dick is Ishmael.

HOW DO WHALES SLEEP?

Like all animals, whales need sleep in order to survive. But how can whales sleep without drowning if they spend their entire lives in water? Whales have solved the problem of simultaneous breathing and sleeping by adopting unihemispheric sleep. This means that one half of the whale's brain is active, while the other half rests. Whales can even close one eye while the other remains open. Whales enter a deeper form of sleep at night but sleep for no more than a few hours or so at most and more typically 30 minutes or less. This behavior is more appropriately known as resting rather than sleeping. Some whales sleep horizontally whereas others, such as sperm whales (*Physeter macrocephalus*), enter a deeper sleep where they hang motionless in groups, vertically not too far below the surface for an hour or so before they surface to breathe. These whales appear to be almost "standing" in the water and tend to cluster in groups of between four and five whales.

↓ Common bottlenose dolphin (*Tursiops truncatus*) sleeping or more likely resting motionless, close to the surface.

→ A group of sperm whales (*Physeter macrocephalus*) sleeping suspended vertically just below the surface. They have been shown to use one half of their brain when sleeping, a behavior that may help them avoid predators, maintain social contact, or control breathing.

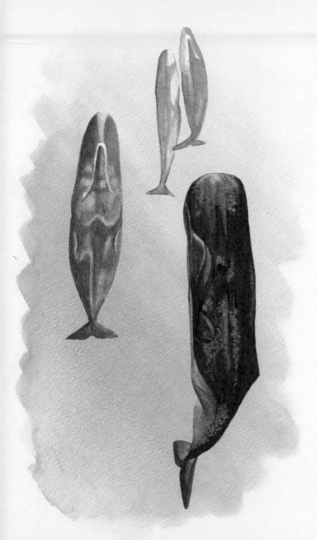

MASS STRANDINGS

Mass strandings occur in many parts of the world when otherwise healthy groups of distressed whales come ashore, generally at the same time and place, strand themselves, and usually die on beaches. These strandings can involve two individuals (not including mother–calf pairs) or hundreds of animals. Both small and large whales are known to strand.

Mass strandings frequently involve social odontocetes, including long- and short-finned pilot whales (*Globicephala melas* and *G. macrorynchus*, respectively); various beaked whale species, notably Cuvier's beaked whale (*Ziphius cavirostris*) and Stejneger's

beaked whale (*Mesoplodon stejnegeri*); false killer whales (*Pseudorca crassidens*); Atlantic white-sided dolphins (*Lagenorhynchus acutus*); and sperm whales. Given their high degree of sociality, when one member of a pod becomes stranded, others often follow.

WHAT CAUSES STRANDINGS?

There are numerous causes of mass strandings, both natural and those resulting from human activity. Some strandings in the past few decades in the Canary Islands, Bahamas, and Hawai'i have been related to military sonar, which may have caused temporary deafness, resulting in whales becoming confused and disoriented.

Many whales that strand show evidence of other physical trauma, including bleeding in the brain, ears, and other tissues. After pathological examination of Cuvier's beaked whales stranded in the Canary Islands during a naval exercise in 2002, decompression-like sickness was reported for the first time, evidenced by gas bubbles, associated lesions, and fat embolisms in vessels and organs. Decompression sickness or "the bends" is a condition best known from scuba divers who resurface too quickly after a deep dive (see Chapter 4, page 57). This suggests that these whales may be susceptible to gas bubble formation in the presence of intense sound pressures, or the sounds may cause them to pursue risky diving behaviors, either of which may make them vulnerable to stranding. Similar pathology was reported following military sonar events from recent strandings (2004–2014) of Cuvier's beaked whales in the Mediterranean.

In other cases, mass strandings are likely caused by weather conditions, diseases such as parasites, toxic algal blooms, and viruses, and changes in underwater topography. Though tragic, these strandings have provided researchers with valuable data on many species that live far offshore in deep waters, thus helping us to better understand and protect these hard-to-study whales.

← A mass stranding of long-finned pilot whales. This species is especially vulnerable given that they live in highly social matriarchal societies.

THE BIGGEST BRAIN

The brain of the sperm whale is five times larger than that of a human and they are distinguished in having the biggest brains of any animal species. Large size is a feature of some whale brains. Brain size is usually expressed with respect to body size, the encephalization quotient or EQ, and has been used as a measure of brain evolution. Social odontocetes, specifically delphinids such as common bottlenose dolphin (*Tursiops truncatus*), killer whales (*Orcinus orca*), and sperm whales, experience significantly higher rates of brain size evolution.

Although the question of how brain size relates to intelligence is controversial, a number of factors drive brain size in whales, including echolocation, complexity of feeding strategy, and perhaps most importantly, degree of sociality. Research on the brains of extinct whales has shown that large brains arose before echolocation, supporting the idea that there were likely several paths to a bigger brain in whales.

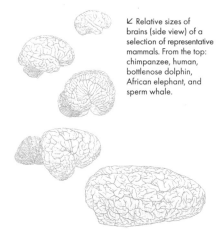

↙ Relative sizes of brains (side view) of a selection of representative mammals. From the top: chimpanzee, human, bottlenose dolphin, African elephant, and sperm whale.

→ The rough-toothed dolphin (*Steno bredanensis*) is distinguished in having relatively large flippers, prominent dorsal fin, and three-toned coloration with a dark gray back, narrow dorsal cape, and belly blotched with white patches. Among cetaceans, dolphins have the highest EQs.

SPRING-ASSISTED SWIMMING

Whales swim by dorsoventral undulations of their flukes, powered by massive axial or trunk muscles. This motion is enhanced by the spring-like elasticity of the fibrous connective tissue in the blubber layer surrounding the tail stalk and flank. When the muscles bend the tail for a downstroke, for example, the elastic rebound of this "blubber spring" will passively bring the tail back up.

Every spring or pendulum has a period—the time it takes to oscillate through one cycle. If you have pushed a child on a swing, you know that if you push in time with the natural period of the swing, it is fairly effortless, but if you push at the wrong time, it is jarring and energy-intensive. Interestingly, the natural period of the blubber spring on the tail stalk of a whale is equivalent to the tail beat frequency associated with its most energy-efficient swimming speed (used for standard travel or migrations). Thus, much of the up-and-down motion of the tail while swimming each day is powered by the elastic oscillation of the blubber spring rather than muscular exertion. Such a simple and efficient mechanical solution beautifully illustrates the genius of nature.

SPEED

The swimming speeds of whales are considerably faster than the average cruising speed for humans, which is 3 ft (1 m) per second. The average swimming speed for both mysticetes and odontocetes is 4.2–11.8 ft (1.3–3.6 m) per second. Dolphins possess the fastest swimming speeds and they can sprint for short distances at over 33 ft (10 m) per second. Whales, especially dolphins, conserve energy by gliding rather than swimming at depth. This switching of locomotion, from passive gliding without any effort or body movement to active swimming (continuously fluking), is similar to terrestrial animals switching between gaits (a horse trot versus a gallop).

EFFICIENT SWIMMING

To help minimize drag and energetic costs such as occur when they surface, whales have developed several unique behaviors to allow them to breathe as they swim. Rather than stroking continuously, they leap into the air and avoid the drag that occurs when swimming at the water's surface. It is also possible that wave riding is a form of play behavior and animals engage in it for fun.

~ Porpoising ~

Small odontocetes, including dolphins (family Delphinidae) and porpoises (family Phocoenidae), moving at high speed, position themselves near the water's surface, leap above the water, and glide airborne for several body lengths. This behavior is known as porpoising.

BOW RIDING

Another strategy used by whales to avoid continuous fluking while swimming is bow or wave riding, best described as surfing the bow or stern wave of a boat (shown below). The bow creates a pressure wave that the dolphin uses to propel it forward. In one report, a pair of common bottlenose dolphins was observed bow riding an 85-ft (nearly 26-m) long blue whale (*Balaenoptera musculus*).

AMBERGRIS

Ambergris was once regarded as the most valuable of whale products and is sometimes referred to as "floating gold." Sperm whales eat large quantities of squid and cuttlefish. Undigested squid beaks (mouthparts) accumulate in the stomach chambers of sperm whales to form a large, glittery mass known as ambergris. Reportedly, it is also thought that ambergris is produced in smaller amounts in pygmy sperm whales (*Kogia breviceps*) and dwarf sperm whales (*K. sima*). A 1993 study reported that the stomach contents of 17 sperm whales yielded only 16 fish but the half-digested remains of 29,000 squids—a prodigious number. The waxy mass is expelled from the sperm whale's mouth every few days and washes up on the shore.

HISTORY

Fossilized evidence of ambergris dates back 2 million years. Ambergris has been traded since the Middle Ages. In the 12th century, reports from China suggested ambergris was dragon spit. An encyclopedia of herbal medicine published in 1491 theorized that ambergris was tree sap, a type of sea foam, or a kind of fungus.

Through history ambergris has been used as a fixative in the making of perfume. Since then, scientists have developed a synthetic version and most perfumes today rely on the laboratory-produced alternative. Ambergris has also been consumed as a delicacy and administered as medicine.

Ambergris has also featured in literature, including in the classic novel *Moby Dick* (see Chapter 11, page 141), with its author, Herman Melville, devoting an entire chapter to this material and writing of the terrible odor from sperm whales, from which "stole a faint stream of perfume."

DNA PRESERVATION

Recent study of the DNA of ambergris has revealed the whale's genetics as well as the whale's gut microbiome and potentially the DNA of its prey. This has the potential to yield valuable information about sperm whale ecology. The preservation of ancient DNA in ambergris could also help scientists estimate sperm whale population sizes before they were pushed to near extinction due to 19th-century commercial whaling.

↑ Ambergris, prized as a fixative in perfumes, is a by-product of the sperm whale's diet. It is composed mostly of undigested squid beaks and accumulates in the whale's digestive system.

SOUND AS A WEAPON

Whales can make the loudest sounds of any animal. Recordings of the world's largest toothed whales, sperm whales, and various beaked whales, such as Cuvier's beaked whale and dolphin species like the Atlantic spotted dolphin (*Stenella frontalis*), reveals that pops or bangs are low frequencies, loud impulse sounds (more than 200 decibels (dB) a measure of sound pressure) that it is thought are used to stun, debilitate, or even kill their prey. For comparison, the normal human range for human hearing is 20 Hz–20 kHz, with hearing damage from sounds louder than 120–130 dB.

FINDING FOOD

Evidence for how sperm whales actually find their food has been scarce despite considerable information about their diet, which is mostly composed of squid, octopus, and fish found in deep ocean waters. Sperm whales are quite formidable, with their giant rectangular heads and large, sound-generating noses that focus pulses of sound on prey. They lack teeth on their upper jaw, so they are unable to bite prey. Instead, they ingest and swallow prey using suction. Several hypotheses for how sperm whales find food are not based on sound. One hypothesis proposes that sperm whales swim upside down using the lighter surface waters to silhouette their prey. Another idea is that they use their bright white jaw line, which comes from the luminescent jelly of squid, to light up their mouths and attract prey.

↘ Skull of an adult male strap-toothed beaked whale (*Mesoplodon layardii*) showing two tusk-like functional teeth in the lower jaw used for dominance battles that result in mating access to females, rather than in feeding.

~ "Big Bang" hypothesis ~

Perhaps best known is the "big bang" hypothesis that originated with whale biologist Ken Norris in the early 1960s. In this hypothesis, which was backed up by some anatomical evidence, sperm whales use loud sounds to debilitate their prey. For example, a large mismatch between successful prey capture (based on examination of stomach contents) and the feeding structures of some odontocetes, such as the absence of functional teeth in most beaked whales and juvenile sperm whales, has been reported. Also, the fact that these whales successfully capture fast, slippery squid and fish without functional teeth suggests that they are able to get very close to their prey before engulfing them. Another example is sucker-marked battle scars across the skin of sperm whales and the presence of gigantic tentacles in their stomachs.

~ Not conclusive ~

Other lines of evidence, however, are not supportive of the "big bang" hypothesis. Acoustic evidence is difficult to collect because sounds of sufficient pressure to damage or disorient fish typically have not been recorded under natural conditions. Experimental laboratory tests revealed no measurable change in behavior for fish (herring, cod, and sea bass, which are some of the favorite food of dolphins) subjected to different acoustic signals.

USING SOUND TO FIND PREY

Research on free-ranging sperm whales found that, although they produce loud sounds, up to 230 dB, they do not debilitate prey acoustically but employ a series of fast repetitive buzzes or clicks to provide high-resolution information when echolocating during active prey chases. In other words, sperm whales use sound to find prey, not to stun it. Although in the future new evidence may show that dolphins are capable of producing high-intensity sounds that stun prey, neither experimental nor field observations thus far support prey stunning among whales in the wild.

GLOSSARY

aquatic nursing strategy
A maternal care strategy characteristic of odontocetes in which the mothers forage during nursing.

anthropogenic
Originating or caused by humans.

apex predator
A predator at the top of the food chain.

auditory bulla
Cetacean "ear bone," comprised of the tympanic and periotic bones.

biomagnification
Series of processes in an ecosystem in which a substance increases in concentration with each higher trophic level.

bow riding
An energy efficient form of locomotion in which whales and dolphins ride the front or bow wave of a vessel.

bradycardia
Slowing of the heart rate, which slows oxygen consumption during cetacean dives.

breaching
Refers to whales leaping out of the water.

bubble net feeding
A feeding strategy in which whales emit a series of bubbles that form a net and trap prey.

capital breeding
A strategy that involves the use of stored resources.

clans
A large grouping of family units in sperm whales.

codas
Pattern of echolocation sound clicks used by sperm whales.

corpus albicans
A permanent scar on the ovary of whales that records the number of ovulations.

counter-current heat exchange
Circulatory adaptation where arteries and veins are adjacent and flowing in opposite directions such that warm arterial blood heats up cool venous blood.

dialect
A group-specific variation in a vocalization that is shared by multiple groups, analogous to different accents applied to the same word.

echolocation click
A short-duration, wide-frequency range vocalization that initiates echolocation.

encephalization quotient
A numeric comparison that considers brain size relative to body size.

gill nets
A wall of netting that hangs in the water and traps dolphins and other marine mammals in addition to targeted fish.

gulper
Feeding technique of a rorqual whale, taking a large gulp of water and filtering it through baleen as it exits.

heterodont
Possessing teeth of different shapes and purposes.

homodont
Possessing all the same teeth with no specialization.

krill
Shrimp-like crustaceans that are the primary food for baleen whales.

k-selected
Refers to species whose populations fluctuate at or near the carrying capacity

of the environment, and which is characterized by long life spans, low mortality rates, large body sizes, and production of few offspring.

lunge feeding
Feeding technique practiced by gulpers/rorqual whales in which they lunge forward, open their mouth wide, and expand their throat to sample a large volume of water.

mate-guard
A form of sexual coercion where males use aggression to control females.

matriline
Descendants of a female.

melon
Fatty forehead of an odontocete, focuses sound for echolocaiton.

myoglobin
Oxygen-binding molecule in the muscles—does not circulate in the blood.

peripheral vasoconstriction
Part of the marine mammal dive response. Vessels constrict, preventing blood flow to the muscles and appendages.

phonic lips
Area of the nasal passage, adjacent to the melon, where odontocetes produce sound.

photo identification
Technique that involves the collection and use of photographs of diagnostic features of whales and other marine mammals for identification purposes.

pingers
Sound devices that are used to prevent marine mammals from net entanglements.

polygynous mating
A mating strategy in which a males mates with more than one female during a single breeding season.

porpoising
Refers to low leaps at the surface of the water made by some cetaceans, notably porpoises.

precocial
Born in an advanced state of development.

primary escort
The dominant male that attends and mates with a female.

promiscuous mating
Mating system in which males randomly mate with females.

rorqual
Mysticetes of the family Balaenopteridae, with expandable throat pleats for lunge feeding.

rostrum
The snout-like extension of cetaceans.

secondary escort
Other "satellite" males near the primary escort that attempt to mate with a female.

skimmer
Feeding technique of right and bowhead whales, swimming continuously with an open mouth and long baleen to filter out slow-moving zooplankton.

sperm competition
Mating strategy seen in some mysticetes in which males attempt to displace or dilute the sperm of other males in order to increase the probability of being the male to fertilize that female.

spermaceti oil
Waxy oil highly prized by whalers that fills the spermaceti organ in the head of sperm whales.

telescoping
Evolutionary changes to the skull by early odontocetes and mysticetes, resulting in elongation and overlapping of skull bones and movement of the blowhole to its current position in cetaceans.

tympanic bulla
Cetacean "ear bone," synonym for auditory bulla.

unit
A social grouping of sperm whales consisting of females and their offspring.

FURTHER READING

Chapter 1: Berta, A., Sumich, J.L. and Kovacs, K.M. 2015. *Marine Mammals: Evolutionary Biology*, 3rd Ed. Academic Press

Gingerich, P.D., Haq, M. u., Zalmout, I.S., Khan, I.H. and Malkani, M.S. 2001. Origin of whales from early artiodactyls: hands and feet of Eocene Protocetidae from Pakistan. *Science* 293: 2239-2242.

Thewissen, J.G.M., Cooper, L.N., George, J.C. and Bajpai, S. 2009. From land to water: the origin of Whales, Dolphins, and Porpoises. *Evo Edu Outreach* 2:272-288.

Chapter 2: Committee on Taxonomy. 2023. List of marine mammal species and subspecies. Society for Marine Mammalogy, www. marinemammalscience. org, consulted on April 17, 2024.

Crossman, C.A., Taylor, E.B. and Barrett-Lennard, L.G. 2016. Hybridization in the Cetacea: widespread occurrence and associated morphological, behavioral, and ecological factors. *Ecol. and Evol.* 6(5): 1293-1303.

Reynolds, J.E., 1999. *Biology of marine mammals.* Smithsonian Institution.

Chapter 3: Wursig, B., Thewissen, J.G.M. and Kovacs, K.M. 2017. *Encyclopedia of Marine Mammals*, 3rd ed., Academic Press, CA.

Chapter 4: Heide-Jørgensen, M.P., Laidre, K.L., Nielsen, N.H. et al. Winter and spring diving behavior of bowhead whales relative to prey. *Anim Biotelemetry* 1, 15 (2013).

Hochscheid, S., 2014. Why we mind sea turtles' underwater business: A review on the study of diving behavior. *Journal of Experimental Marine Biology and Ecology*, 450, pp.118-136.

Ponganis, P.J. (2015). Diving physiology of marine mammals and seabirds. Cambridge University Press.

Quick, N.J., Cioffi, W.R., Shearer, J.M., Fahlman, A. and Read, A.J., 2020. Extreme diving in mammals: first estimates of behavioural aerobic dive limits in Cuvier's beaked whales. *Journal of Exp. Biol.* 223(18).

West, K.L., Walker, W.A., Baird, R.W., Mead, J.G. and Collins, P.W., 2017. Diet of Cuvier's beaked whales *Ziphius cavirostris* from the North Pacific and a comparison with their diet world-wide. *Marine Ecology Progress Series*, 574, pp.227-242.

Chapter 5: Farnkopf, I.C., George, J.C., Kishida, T., Hillmann, D.J., Suydam, R.S. and Thewissen, J.G.M., 2022. Olfactory epithelium and ontogeny of the nasal chambers in the bowhead whale (*Balaena mysticetus*). *The Anatomical Record*, 305(3), pp.643-667.

Feng, P., Zheng, J., Rossiter, S.J., Wang, D. and Zhao, H., 2014. Massive losses of taste receptor genes in toothed and baleen whales. *Genome biology and evolution*, 6(6), pp.1254-1265.

Hüttner, T., von Fersen, L., Miersch, L. and Dehnhardt, G., 2023. Passive electroreception in bottlenose dolphins (*Tursiops truncatus*): implication for micro-and large-scale orientation. *Journal of Experimental Biology*, 226(22).

Chapter 6: Bezamat, C., Simões-Lopes, P.C., Castilho, P.V. and Daura-Jorge, F.G., 2019. The influence of cooperative foraging with fishermen on the dynamics of a bottlenose dolphin population. *Marine Mammal Science*, 35(3), pp.825-842.

Bowen, W.D., 1997. Role of marine mammals in aquatic ecosystems. *Marine Ecology Progress Series*, 158, pp.267-274.

Clarke, M.R., 1996. Cephalopods as prey. III. Cetaceans. *Philosophical Transactions of the Royal Society of London. Series B: Biological Sciences*, 351(1343), pp.1053-1065.

Connor, R.C. and Peterson, D.M., 1994. *The Lives of Whales and Dolphins*. Henry Holt and Co, NY.

Kaplan, J.D., Goodrich, S.Y., Melillo-Sweeting, K. and Reiss, D., 2019. Behavioural laterality in foraging bottlenose dolphins (*Tursiops truncatus*). R. Soc. Open Sci, 6, 190929.

Savoca, M.S., Czapanskiy, M.F., Kahane-Rapport, S.R., Gough, W.T., Fahlbusch, J.A., Bierlich, K.C., Segre, P.S., Di Clemente, J., Penry, G.S., Wiley, D.N. and Calambokidis, J., 2021. Baleen whale prey consumption based on high-resolution foraging measurements. *Nature*, 599(7883), pp.85-90.

Tucker, M.A. and Rogers, T.L., 2014. Examining predator–prey body size, trophic level and body mass across marine and terrestrial mammals. *Proceedings of the Royal Society B: Biological Sciences*, 281(1797), p.20142103.

Wursig, B. 2019. *Ethology and Behavioral Ecology of Odontocetes*, Springer, CH.

Chapter 7: Henry, L., Barbu, S., Lemasson, A., & Hausberger, M. (2015). Dialects in animals: Evidence, development and potential functions. *Animal Behavior and Cognition*, 2(2), 132-155.

Janik, V.M., Sayigh, L.S. Communication in bottlenose dolphins: 50 years of signature whistle research. *Journal of Comparative Physiology A* 199, 479–489 (2013).

Chapter 8: Jefferson, T. A., Webber, M. and Pitman, R.L. 2015. *Marine Mammals of the World: A Comprehensive Guide to their Species Identification*, Academic Press, CA.

Mann, J., O'Connor, R.M., Tyack, P.L. et al., 2000. *Cetacean Societies: Field Studies of Dolphins and Whales*, University of Chicago Press.

Chapter 9: Berta, A. 2012. *Return to the Sea*, University of California Press, CA

Berta, A. 2015. *Whales, Dolphins and Porpoises: A Natural History and Species Guide*, University of Chicago Press, IL.

Berta, A. 2023. *Sea Mammals: The Past and Present Lives of Our Oceans' Cornerstone Species*, Princeton University Press, NJ.

Clark, C.W. and Garland, E.C. 2022. *Ethology and Behavioral Ecology of Mysticetes*, Springer, CH.

Swartz, S.L. 2014. *Lagoon time: a Guide to Gray Whales and the Natural History of San Ignacio Lagoon*, Ocean Foundation (Sunbelt), CA.

Chapter 10: Braulik, G.T., Taylor, B.L., Minton, G. et al. 2023. Red-list status and extinction risk of the worlds' whales, dolphins and porpoises. *Conservation Biology* 3:37:e14090.

Davidson, A.D., Boyer, A.G. Kim, H. et al. 2012. Drivers and hotspots of extinction risk in marine mammals. *PNAS* 109(9): 3395-3400.

Ellis, R. 1991. *Men and Whales*. Alfred A. Knopf, NY.

Moore, M.J. 2021. *We Are All Whalers: The Plight of Whales and Our Own Responsibility*, University of Chicago Press, IL.

Chapter 11: Heithaus, M.R., 2001. Predator–prey and competitive interactions between sharks (order Selachii) and dolphins (suborder Odontoceti): a review. *Journal of Zoology*, 253(1), pp.53-68.

Mann, J. (ed), 2017. *Deep Thinkers: the Mind of Whales, Dolphins and Porpoises*, University of Chicago Press, IL.

Pyenson, N.D. 2017. *Spying on Whales*. Viking, NY.

Chapter 12: Catton, C. 1995. *Dolphins*, St. Martin's Press, NY.

INDEX

ACKNOWLEDGMENTS

We are pleased to acknowledge the highly capable editorial team at UniPress, especially Ruth Patrick, project development and management, Lindsey Johns, design and art direction, Caroline West, copy editor, Slav Todorov, managing editor, and creative director, Alex Coco. We also thank Tugce Okay and Ian Dureen for their expertly drawn illustrations, and Nature Picture Library and Alamy Stock Photos.

ABOUT THE AUTHORS

Robert Young is a Professor of Marine Science and the Associate Provost for Research at Coastal Carolina University, where he has taught courses in marine science, marine biology, marine mammals, and fisheries science for 32 years. He served as coordinator for the South Carolina Marine Mammal Stranding Network for 14 years and is a past President of the South Carolina Marine Educators Association. His research and publications have focused on the ecology and behavior of coastal and estuarine dolphins and fish, with particular attention to the ecological bioenergetics, behavior, and stock structure of estuarine and coastal bottlenose dolphins in the southeast US.

Annalisa Berta is Professor Emerita of Biology at San Diego State University. A specialist in the anatomy and evolutionary biology of marine mammals, especially baleen whales, she formally described a skeleton of the early pinniped *Enaliarctos*. She is the author of *Return to the Sea: The Life and Evolutionary Times of Marine Mammals* and *Sea Mammals: The Past and Present Lives of our Oceans' Cornerstone Species*, as well as the editor of the award-winning *Whales, Dolphins, and Porpoises: A Natural History and Species Guide*.